SOIL IN THEIR SOULS

A HISTORY OF FENLAND FARMING

REX SLY

The History Press

Also by Rex Sly

From Punt to Plough: A History of the Fens
Fenland Families

First published 2010
Reprinted 2018, 2024

The History Press
97 St George's Place,
Cheltenham, Gloucestershire, GL50 3QB
www.thehistorypress.co.uk

British Library Cataloguing in Publication Data.
A catalogue record for this book is available from the British Library.

ISBN 978 0 7524 5733 8

Typesetting and origination by The History Press
Printed by TJ Books Limited, Padstow, Cornwall

Contents

Acknowledgements

I am deeply indebted to all who have helped me in writing this book. There are those mentioned in various chapters whose time and input have enabled me to complete this documentation of fenland farming. They have often given me their time when it is was precious in this industry, as well as their valuable knowledge and an understanding of what I was striving for. So much has been lost, or misplaced in records of fenland agriculture and so it was a joy to unearth some of what is in the book. I thank all who have delved deep in their cupboards for books, ledgers and photos of their families for me to use, much of which they themselves had forgotten they had, or did not even know was there.

Many people who are not directly involved in agriculture have helped me to gather information, in libraries, museums, and offices, some of whom are connected with this industry. Some of those people may have thought their input at times was irrelevant, but to me it was vital to be able to find out not only when changes occurred in our industry, but why.

There were people who have helped me who are not mentioned in the book; to them I am indebted and would like to say, without them it would not have happened. Once again, for three years Maggie has patiently hovered around me when writing, she can now give my office a thorough cleaning ready for the next crop, whatever it may be. More than anyone I thank my wife Steph for her patience, tolerance and encouragement – qualities without which I would never have completed my third book.

Introduction

The past century has brought to bear untold pressures on our soils as we endeavour to feed our nation while also bowing to government agricultural policies. Many of those involved both within and outside the farming industry during this period did not realise the consequences of over-production, and still today the burden on our soils to produce crops continues to be fuelled by the ever-increasing demands from the retail sector.

Nowhere is the impact of this more evident than in the lands surrounding the Wash – the marshes and their hinterlands, the Fens. This land is about soil and the people who manage it. The land has been drained so that farmers, graziers and growers from each generation could earn their living from the soil while at the same time helping to feed the nation. Its people have, over centuries, extended their domain by reclaiming land from the sea, using sheer manpower as well modern machines.

The custodians of this precious area we call the Fens live in scattered parts of the fen and marsh. Some live on the farms, some in towns and villages, and many down at the bottom of long fen droves, gates, roads and banks. The farmers in the marsh are truly 'Marsh men', and those in the fen 'Fenmen'.

Small farmers and growers have by sheer hard graft and tenacity survived in an ever-changing world where the 'winner takes all', but their marketplace has changed as little over time as they themselves have. Alongside them are the large hi-tech agri-businesses and growers capable of fulfilling the requirements and demands of supermarkets and processors, a world driven by the marketplace. In many ways, the two farming sectors are decades apart but they have a common bond: the soil. It is the unique soils of this man-made farming region that have provided, and still do provide, a living for both sectors of agriculture and horticulture.

Change has been part of the Fens' history ever since man first set foot there. The Fens have always been a land of adventurers, of drainers and engineers, of land speculators along with their agents. It has attracted farmers and growers from home and abroad, alongside packers and processors of fruit vegetables and flowers of every description, where innovators and traditionalists work side by side.

The Fens were the birthplace of the allotment and small-holders' movement over a century ago which gave men the chance to step onto the farming ladder and provide a better life for their families. The Fens' labour force has been sourced from around the world by the lure of the lucrative work this area has had to offer – although not all by choice! POWs from the two world wars were among that number in the early days of draining. In addition, students from around the globe have trodden the fenland soils in pursuit of work experience and pocket money.

The drainage of the Fens has, over the centuries, constantly kept in unison with the demands of agriculture, its industries and its inhabitants. Its soils have changed, as have farming practices with each generation, which in turn has had an effect on the environment. The farmers and growers understand these changes and the challenges they present – they are second nature to them. More than anyone they know, that 'our soils are our lifeline for the future' (a quote from Archie Saul mentioned in this book).

These soils have been wedded to the human footprint since man first claimed them from nature. Not footprints left by the rambler and traveller, but the agriculturalists and labourers of this land. The natural richness of its soils and the crops grown required an abundance of hand labour and horse power. Much of the fenland cropping was intensive involving row-cropping. Crops such as peas, potatoes, sugar beet, onions, vegetables and many other root crops were sown by horse-drawn drills and planters in rows. Fertilisers were applied with horse-drawn spreaders, and reapers were pulled by horses, led by a labourer. Each operation left a human and equine imprint in the soil, a seal of contact. The horses' imprint died a natural death in the mid-twentieth century to be replaced by the tyre and the track, and soon after modern agronomy and machines replaced the human footprint. There are still human footprints on some areas of the soil, mainly from flower cropping, vegetable and salad harvesting, but it is a tiptoe of former days.

On my journey through the Fens I visited some of the people involved in the farming industry. Many I knew, as you would expect considering as I have lived and worked among them all my life, and those I did not know, I understood because of a common bond we have: 'Soil in our Souls'.

Rex Sly, 2010

1

The History of Fenland Agriculture

We know that the Fens started to become much drier from about 700 BC from the number of Iron Age settlements that have been found to have existed across the entire area. The Roman period of occupation was from AD 40 to AD 650 and their settlements were also scattered across the Fens. Small-scale agriculture was evident in these settlements, but salt production was most prolific. Human settlements from these periods were not always found on the higher silt or gravel islands – many were on what we now refer to as the lower fen – leading us to believe that either sea levels were lower then or water from the surrounding highlands draining into the Fens was not as great as at other periods in history. It could also have been a period of less rainfall, maybe owing to climate change or just a drier cycle of years.

Some 200 years after Roman occupation, Celtic monks began to settle mainly on the fen edges and on isolated islands in the Fens themselves. Abbeys were established near the fen edge at Peterborough, Ely and Ramsey while Crowland, Thorney and the nunnery at Chatteris were furthest in the Fens. A religious settlement at Algarkirk was on the edge of the marsh. Agriculture at this period would have been pastoral, mainly cattle and sheep, with banks as defences being the method of trying to keep at bay the inundations from the sea.

Tillage was well established across the UK but was used only on the small areas of higher ground in the Fens. Flooding from the sea and water from the highland run-off favoured pastoral farming, and livestock could be raised and fattened on summer grazing in the fen then moved to the surrounding higher ground in the winter. Land bordering the villages and towns in the Fens supported this method of minor arable farming.

During the period up to the Reformation in the sixteenth century, much of the fen and marsh was either owned by the religious orders, under the jurisdiction of the abbeys or priories, or by the ancestors of the Norman families. These areas were mainly associated with the manors, granges and small religious cells. After the Reformation, land ownership fragmented and little was done to drain land.

By now the UK had extended its colonies around the globe, serviced by its merchant fleet and patrolled by its large navy. The global market, a phrase we so often apply to the twenty-first century, was already in place. The Hanseatic League, a powerful corporate body

Crowland Abbey, built on an island of gravel.

formed in the thirteenth century by a group of northern German trading towns, was the first great trading organisation in Europe. It dominated the main trade in northern Europe and incited several trade wars, until its demise in 1669. 'The Hanse' had warehouses in two fenland ports: King's Lynn and Boston. I have found no mention of the commodities they traded in and out of these ports, but we do know, however, that wool was sent to the Flemish weavers from this area and the import of Port wine from Portugal became established by families such as Yetman of Boston in 1708 and Elgoods of Wisbech.

The seventeenth century saw the birth of fenland agriculture as we know it today, brought about by the drainage schemes of the four great levels of the Fens. This was the age of speculators and adventurers. The British were establishing colonies around the world and investing large sums of money to develop them. Britain had recovered from the Black Death, and the growth in population was outstripping food supply. Many wealthy individuals from the City of London, courtiers and other aristocratic families had seen what the draining of the Isle of Axholme Level by Dutch speculators had achieved and wished to follow suit.

There were around 750,000 acres in the Fens, of which almost 500,000 acres were there for the taking: untapped virgin land waiting to be drained. If that could be achieved it would create another province of England. After the Reformation only three of the fifteen abbeys remained in the Fens. Some courtiers did obtain the abbey estates but most of that land was either common land or in the ownership of a very few owners and populated by peasants.

The Bedford Level, covering the major part of the central and southern section of the Fens was the largest scheme, covering some 300,000–400,000 acres. Other levels included the Deeping Level, comprising some 30,000 acres between the Rivers Welland and Glen, while the Lindsey Level, which encompassed Holland Fen, lay between the River Glen and River Witham in the north and covered in excess of 50,000 acres. The most northern level was the area between the Rivers Witham and Steeping, an area of about 45,000 acres encompassing East and West Fens as well as Wildmore Fen. The total area is difficult to pinpoint accurately, but must have been about 4–500,000 acres.

Adventurers undertook the financial task of draining these levels and in return, by royal commission, were granted the freehold of designated areas. Where freeholders already existed they retained their land, but had to pay drainage payments when work was completed.

However, at this point almost all this low-lying land was common grazing land, not tillage. It sustained cattle, sheep and pigs as well as providing fishing and fowling for the indigenous population. They also practised communal cropping around the villages and hamlets on mainly copyhold land owned by the relevant lord of the manor. Much of this land would have been drier land and did not come under the great drainage schemes. Due to the land being only for summer grazing, most of the livestock was slaughtered at Michaelmas (29 September), with only breeding stock and draught oxen being over-wintered in and around the villages. Dung from these animals provided excellent manure for their arable plots – what would today be called 'sustainable farming'.

The loss of the common lands affected the indigenous fen people most of all. They relied on it for their very survival, and could see that the loss of it would bring extreme hardship and maybe the end to their way of life. Common pasture was their birthright handed down since their ancestors pioneered these Fens centuries earlier. They had learnt from their forebears how to farm and manage this land in all its moods and accepted their lot with nature. Many disturbances and several riots did occur in protest to the drainage schemes, primarily because of the loss of common grazing but not, as is often claimed, because of the fenmen's opposition to drainage.

A speech delivered by Sir John Maynard, a leasing upholder of the fenmen's cause before the Committee of the Lincolnshire Fens in 1650, contains one of the best summaries of the value of the traditional fenland economy. Complimenting its wonderful pasture lands he says:

> Our Fens as they are, produce great store of wool and lamb, and large fat mutton, besides infinite quantities of butter and cheese, and do breed great store of cattle, and are stocked with horses, mares, and colts, and we send fat beef to markets, which affords hides and tallow, and for corn, the fodder we mow off the Fens in summer, feeds our cattle in winter: By which means we gather such quantities of dung, that it enriches our uplands, and corn-ground . . . Besides, our Fen relieves our neighbours, and uplanders, in a dry summer, and many adjacent counties: So thousands of cattle besides our own are preserved, which otherwise would perish . . . So that Rape, Cole-seed and Hemp, is a Dutch Commodity, and but trash and trumpery, and pils Land in respect of the afore recited Commodities, which are the Oar of the Commonwealth.

Sir John Maynard was a leading Presbyterian who died in 1658. Quoted from (*Fenland Riots and the English Revolution* by Keith Lindley, page 7).

One contemporary estimate quotes:

> nearly 4,000 families residing in the Skirbeck Wapentake of Holland and the Soke of Bolingbroke in Lindsey, were to some extent dependant upon the East and West Fen Commons. Wildmore Fen alone, an area of some 10,000 acres, supported 930 families in the Sokes of Horncastle and Bolingbroke. Eleven villages in Kirton and Skirbeck had common rights in Holland Fen. Many thousands of families in the Wildmore, East and West Fen Level and Holland Fen being part of the Lindsey lost their grazing rights.

This area probably amounted to about 15 per cent of the total area of the levels drained in the seventeenth century and we are led to believe around 5,000 families lost their common land here. If these figures are used across the entire four levels drained, possibly upwards of 30,000 fenlanders in total would have lost their grazing rights – figures that are hard to comprehend.

The meres and wetlands represented for some a way of life that would end once drained, and those with grazing rights were also restricted. Much of the new lands were now for winter as well as summer grazing, and would remain so until nineteenth and twentieth centuries, especially the lower-lying land. The majority, however, did revert to tillage. The great level areas were mostly taken over by large landowners who sub-let to previous grazers and tillage farmers. Smaller enclosures across the Fens were often taken by local landowners and sub-let, with farmers at that time being mainly tenants or copyhold leasers.

The draining of the four levels did go ahead in various stages in the seventeenth century. Not all were successful in their initial stages but it was the period in our Fen history when we could say 'the Fen farmer had been conceived'.

Land was let to farmers by covenants. For example, around Wisbech it was common practice to let land by auction for 20 years by covenant. The covenants included clauses to restrict farmers to three- or four-course rotations specifying which crops could be grown, including fallow. Crops such as hemp, flax, woad, madder, cole seed and mustard seed were deemed to take fertility out of the soil, while turnips fed to stock, tares, beans and peas added to fertility. Paring and burning on the peat fen was also restricted. Some landlords specified what portions of the land could be cropped and what had to be in grass. The restrictions by landlords gave little incentive for tenants to bury their souls into the soil for long-term agriculture.

The use of artificial fertilisers was not yet common practice. Grazing of sheep on crops such as turnips, and the use of grass leys and farmyard manure were the only ways of maintaining fertility in the soil. All these practices required more capital than arable farming for the tenants. Guano fertiliser from Peru was not introduced to the UK until the late nineteenth century. Large areas of the poorly drained Fens at this time were either left to permanent grass, which was mostly couch grass, or some of it would be ploughed, pared and burned then arable cropped until the fertility was exhausted, after which it was left to revert to its original state of couch grass and weeds.

Agriculture in the Fens changed little over the eighteenth and early nineteenth centuries, much of it remaining a combination of mainly pastoral farming, with most arable farming taking place on the silt soils of the higher ground. The livestock farming would have been mainly fattening cattle on grass during the summer months, not breeding herds. The land was too wet during the winter months for grazing and there would have been insufficient winter keep to yard cattle for the winter. However, by the mid-nineteenth century this was changing as many of the larger landlords started to build agricultural farm buildings to cater for the abundance of straw crops and roots and to yard and feed cattle during the winter months.

Indeed, fattening store cattle in the Fens became big business with the advent of potato growing, feeding waste, and more availability of straw for bedding in the period from the mid-nineteenth century to the 1960s. The Fens have always been reliant on store cattle from the all parts of the UK, and southern Ireland during the post Second World War years.

Grants and subsidies for farmers are frowned on by much of the population today and thought to be a modern conception, yet this is far from the truth. This precarious industry that is so vital for man's existence has for so long been supported through perilous times. As far back as the seventeenth century, governments had supported agriculture partly due to the dismal state of farming. A letter from Mr John Maynard of Holbeach Marsh in 1653 mentioned that 'corn is cheap and money scarce,' while tenants from a Norfolk estate pleaded for the forbearance of their landlords in paying their rents since 'neither corn nor cattle nor butter nor cheese will give any price.' From Oxford in May 1653 it was reported that a bushel of wheat, which some years before costs 10s was being sold for 2s 6d or 3s. The population of an island nation such as Britain was controlled by disease and its food supply. Food supply was determined by weather, alongside animal and crop diseases.

Government help supporting grain farmers with bounties on exports continued well into the eighteenth century. The government was worried the lack of profitability for the grain farmer would drive him into alternative crops such as grazing livestock. This was a risk they could not afford to take with grain being the main feed commodity for man, beast and the brewer. Livestock farming employed far less labour than arable and unemployment would be of concern in rural communities where no alternative employment was available. As a consequence, livestock farmers were less fortunate in gaining support but were compensated during the Great Cattle Plague of 1714 and 1745–59.

The farming fraternity will associate the eighteenth century with names such as Viscount Townsend who lost his position as prime minister in 1730 and returned to his Norfolk Estate to oversee his farming operations. He became know to farmers as 'Turnip Townsend', as he fathered the 'Norfolk Four Course Rotation'. This became the basis for good farming husbandry and would be taught to young farmers for the next 200 years. The words farming husbandry and rotation were coming of age.

In 1745 Robert Bakewell began his programme of livestock breeding with cattle and sheep, which many would emulate to produce the many breeds we have today. Many of these breeds would in the future dominate the prairies and pampas grasses around the globe. In 1776 Thomas Coke, Earl of Leicester, began experimenting in scientific farming practices. Books and manuals were also being published on agricultural methods of production, husbandry and other aspects of farming and land management.

In the General View of Agriculture in the County of Cambridgeshire, 1811, Gooch states that there was 50,000 acres of improved fen, 150,000 acres of waste and

An 1814 bill of sheep sale at Smithfield Market, London, from Tilney All Saints.

Tilney All Saints.				STEPHEN GREGORY.			
Direct to Me at Messrs. Sharpe & Sons,							
West-Smithfield, London.							
104 Sheep sold for Mr. R. Crane 28 of Feb: 1814.							
HM	Howard 112/-	212	16	,,	Selling & Charges	4	6	8
7	Balsh 116/-	40	12	,,	Grass, &c............	—	—	—
5	Neesham ...115/-	28	15	,,	Letters	—	—	—
M7	Howson 116/-	156	12	,,	Carriage	,,	6	8
10	Tuck 112/6	56	5	,,	Drover	8	13	4
1	Bagley 100/-	5	,,	,,	To............	580	1	4
11	Boal'...........118/-	64	18	,,				
5	Hatton 114/-	28	10	,,				
104		£ 593	8	,,		£ 593	8	,,

William Hebb

unimproved fen, and 8,000 acres of fen and moor common. These figures only account for the fenland area of Cambridgeshire and, if correct, it would be fair to assume that at this period around half the entire Fens would have been unproductive for agriculture.

It was the mid-nineteenth century, with the advent of steam power, centrifugal pumps and railways, which was to change the Fens. While the eighteenth century would be remembered for the innovators in farming, both with livestock and arable husbandry, the nineteenth century saw the birth of mechanisation in agriculture. Ploughs, cultivators, drills, reapers and thrashing machines dominated the farming news headlines. The power of steam alongside the horse was turning worthless fen into a farming paradise. Landowners such as the Duke of Bedford spent vast sums on drainage as well as on the most modern farmhouses, cottages and agricultural buildings on his 20,000-acre Thorney Estate. He and other fenland landlords thought nothing could go wrong with British agriculture, even with the repeal of the Corn Laws which were introduced in 1815 to protect the wheat prices. Little did they know what was in store.

This period was the renaissance of British agricultural buildings. Across the UK landlords were excelling at building farm premises of outstanding design and splendour. There were two such landlords in the Fens, the Duke of Bedford at Thorney, and the Lord Exeter in Postland and Deeping Fen, both of whom built farmsteads on their estates, few of which remain today.

Among the most important acts of government during the nineteenth century affecting the Fens were the Enclosure Acts, not only the acts to enclose land, but the General Enclosure Acts, safeguarding tenants' rights. The acts changed the management of common land from being virtually non-rotational, and without any husbandry, to managed farming. The word agronomy was born in 1814, meaning the science of managing farmland, borrowed from the French *agronomie*. The national figure of enclosures between 1760 and 1867 was 7,500,000 acres.

The Royal Society of Agriculture was formed in 1838, but the Fenmen were one jump ahead. Three years before, in 1835/6, Long Sutton Agricultural Association was formed. It comprised a district within a radius of 30 square miles. Long Sutton, or Sutton St Mary as it was known, included the three parochial chapelries and townships of Sutton St Edmunds, Sutton St James

In 1851, Marshland Cut was opened, and the water of the Mere had drained away in three weeks. All went well till November, 1852, when the banks of the outer-drains burst. In a few hours the Mere was itself again, and the throbbing pump was at work raising 20,000 gallons a minute.

The first culture was with a hand harrow. The frost of winter and dry winds of March made the soil so porous that a high wind wafted several inches of the surface away, carrying with it the sprouting wheat of the first crop. To solidify the surface, clay was brought by means of a portable railway from Holme Park. The claying cost £12 an acre. Grubbers and scarifiers were used so as to avoid disturbing the peat. The clay once buried in a furrow gradually subsided to the bottom of the peat, out of the reach of all crops.

L.G.

1851 notes on the draining of Whittlesey Mere.

and Sutton St Nicholas which contained 21,404 acres. This area had been increased considerably over the past half-century with the enclosure of 15,000 acres from the sea on both sides of the River Nene. Many parishes around the Wash had also increased their acreage – Holbeach had increased by 12,390 acres and Whaplode by 1,223. Not only was land being reclaimed from the sea, but also from the fen itself. Whittlesey Mere was drained in the mid-nineteenth century, an area of some 2,000 acres, as well as many other meres in the southern and northern sections of the Fens. This was brought about because of the invention of steam engines driving centrifugal pumps.

Many other low-lying parts of the fen areas previously used for grazing were also being turned into arable land. In 1801 an act of parliament was passed to enclose 34,000 acres in Deeping Fen and surrounding area. This took almost four decades to accomplish – in fact it wasn't finished until the advent of steam-driven pumps. There were many acts during the reign of King George III amounting to several thousand acres across the whole of the Fens, especially in the low-lying areas away from the Wash. Holland Fen had 22,000 acres enclosed at the end of the eighteenth century which was divided and allotted to the surrounding parishes. An area well in excess of 100,000 acres was probably enclosed during this century and being prepared for intensive agriculture.

By the nineteenth century Great Britain was supreme as a nation. Its overseas empire spread around the globe and at home its industrial might was growing at an alarming rate. The advent of steam power, together with the new machinery in the mills, furnaces and factories fuelled by demand, had made this country the leading industrial nation on Earth. There was, however, among the industrialists a feeling of unrest. They wanted cheaper food for the growing population of factory workers, but their hands were tied by the landed gentry.

In the mid-nineteenth century, 7,000 people owned 80 per cent of the land in the UK. By status, if you owned between 100 and 1,000 acres you were considered a yeoman, between 1,000 and 10,000 acres you could be considered gentry. If, however, you owned 10,000 acres or more you were an aristocrat, and this sector of society owned the majority of the land in the UK. It would be due to their influence that the Corn Laws were introduced in 1815 to protect the wheat prices for themselves and the tenants on their estates. The landed gentry still had great influence where agriculture was involved.

There was a large sector of people who had made money in commerce, trade, the professions and industry and become landowners, but they lacked the relationship

In the late nineteenth century a gang of men harvest with scythes.

between landlord and tenant that the great landowning families had. Their financial interests lay more in harmony with where their wealth had materialised from than their agricultural estates. Many were more interested in cheap food for their workers than the lot of their tenant farmers. Mobility had arrived in agriculture for harvesting, thrashing and transport of corn. The death knell had been rung for the small miller and it would be only a matter of time before he would become extinct.

The Corn Laws would become a battleground between the new industrialists and the 'old landed aristocracy'. In 1838 statesmen and leading merchants in Manchester formed an anti-Corn Law association with an aim for free trade in corn and the repeal of the Corn Laws of 1815. They lobbied for eight years. The Irish potato famine of 1845, which affected the staple diet of the Irish peasantry, caused immense hardship in that country and served to bolster their cause. In 1845 the prime minister, Robert Peel, persuaded the government to repeal the Corn Laws, although sanctions on imports were not fully abolished until 1869.

The nineteenth century would become the golden era of the corn industry in the UK, yet ironically it would prove also to be the worst in agricultural history. The Lands Improvement Company of 1853 became the largest lender – by 1880 it had lent more than £4 million.

Modern farming practices, improved drainage, and more land in agricultural production increased the farmers' yields of crops. Steam power available for flour and provender mills, as well as water pumps and thrashing machines setting the scene for a new era in agriculture. Labour had left many of the farms to earn higher wages in the cities and towns, but with new machines on the farms to bind, thresh and handle corn crops, this was not a problem. The 1881 census showed a decline of 92,250 agricultural labourers since 1871. However, the horse was still the main source of power on the farms and remained so until the early part of the twentieth century when they were replaced by tractors.

The growth of railways between 1840 and 1860 ran into thousands of miles of tracks linking areas of the UK. It was especially significant in the Fens, which had always been cut off from the rest of England. By 1850 the Fens had branch lines joining most of the towns and villages which interlinked with the main lines to other parts of the UK. Second to drainage, the railways would change the Fens more than any other event in its history. The fenland markets also witnessed a surge in their trade with the outside world. Over the next three decades corn exchanges were built in every town in the Fens, for farmers and growers to trade with buyers who had travelled from around the UK. Firms from far and wide had representation in the corn exchanges with their representatives travelling in by train – little did they know many of the exchanges would close within a few years of being built.

With the growth in our overseas colonies by the 1870s cheap agricultural goods, especially grain, was flowing into the UK at an alarming rate from around the world. Britain's dependence on imported grain in the 1830s was 2 per cent, by the 1860s it was 24 per cent rising to 45 per cent in 1880. In 1913 the UK was importing 5 million tons of wheat, three times its domestic output. British farmers were being undercut in prices and as a consequence reduced their arable acreage during the latter part of the nineteenth century and during the years prior to the First World War.

Not all farming during this period was in a depressed state. Cheap grain meant cheap cattle feed, hence the livestock farmers and mixed farms saw their incomes rise steadily between the repeal of the Corn Laws and the end of the nineteenth century. Beef, mutton and dairy producers also enjoyed a rise in income. In 1919, 20 per cent of the UK agricultural output was from the dairy sector but the rest was unprofitable. Not all the arable sector suffered as farmers growing potatoes in the fen and marsh, especially on the silt soils bordering the Wash, witnessed good times, many of them making large fortunes in the process.

A meeting of the Wrangle Cow Insurance Club in the mid-1960s. Among those seated around the table are Tom Sinclair, Bill Hanks, Frank Edwards and Peter Marston.

Carting and stacking hay in the early twentieth century.

Gathering in the harvest by punt in 1912.

In 1911 within a 6-mile radius of Wisbech, 5,000 acres were under fruit production growing mainly apples, plums, bush and cane fruit and asparagus, which did not suffer the agricultural depression as much as cereal growers.

Much of the upland farms had turned arable land into pasture or let it go to scrub. This had also happened on the heavier fenland farms. Gold Dike farm in Thorney Fen, consisting of about 300 acres, was sold to the tenant when the Duke of Bedford's estate was broken up in 1909/10. It was made up of 166 acres of pasture and 134 of arable land. Unlike large areas of the Fens at that time, which were poorly drained and down to pasture for that reason, Thorney Fen was relatively well-drained. The plan of the farm at that time shows the fields to be of good size and shape and more than likely they had been arable farmed before the start of the 1880s' agricultural depression.

From the mid-nineteenth century up until the First World War, politicians were prepared to watch British agriculture go into decline with the availability of cheap food from around the world, a sentiment also shared by the British public. A quotation I recall from Canada at this period read, 'A farmer was considered an essential factor in the scheme of things which gave him self-respect, while in England there is an interesting contrast in the attitude of quite a large proportion of the British public today who look upon him as an unnecessary nuisance.' Words still heard in some quarters of society today!

The demise of agriculture drew farmers together as did the workers' unions. The Lincolnshire Farmers' Union was founded in 1905 and the Whittlesey branch formed in 1909 and Ramsey 1911. The National Famers' Union was also founded in 1909. At the outbreak of war, the area of cultivated arable land in the UK had declined dramatically and over half of our food was imported. Even in 1916, the government seemed hardly concerned about British agriculture. No attempt was made to make farming a reserved occupation, as happened in the Second World War, or to restrict horses being commandeered from farms to be used in the Army. When ships bringing food to the UK suffered on account of the German U-boat threat the tide turned in favour of the British farmer.

Guaranteed price controls were introduced which in turn made wheat growing profitable, turning large areas of grassland back to arable. In 1917 the Corn Production Act was passed guaranteeing the wheat price at 60s a quarter and oats at 38s. The Agricultural Wages Board was set up and landlords were restricted on their rent increases and tenants were free to plough grassland, previously restricted in their leases. The war had brought home the importance of British agriculture and changed large areas of the Fens from grazing land to arable.

Farming in the Fens was also helped by the sugar beet crop, making it a major cash-earner in the rotation. The Sugar Industry (Subsidy) Act of 1925 made investment in new sugar factories far less risky and a total of four sugar beet processing factories were built in and on the Fens (see section on sugar beet page 133).

However, another agricultural depression was just around the corner and farming in the Fens like most of the UK went into freefall again in the late 1920s and early '30s. Much arable land reverted back to grass until the late 1930s when the start of another world war loomed on the horizon.

Some of the heavier land in the Fens was being let by landlords to tenants rent-free for two years as an incentive to help them make a living. Many of the large landowners were also either reducing their tenants' rents or, in some cases, waiving them until times

Carrots being trans-shipped from horse and cart to a fen lighter near Chatteris in the early twentieth century.

One of a pair of steam traction engines winching a cultivator across the fen in the late nineteenth century.

An early twentieth-century Spalding sheep market.

improved, rather than have vacant farms. My father said many farmers lost money by farming the land and would have been better leaving it fallow, as in that way their losses would have been reduced. His words remain engrained in my memory, 'During wars the farmer and worker are classed as heroes; between the wars no one wants them.' The advent of war quickly changed this situation again.

Mechanisation on the farms and diesel-driven pumps along with grants to drain the land more efficiently brought about the greatest change in farming since the seventeenth century. By 1936, 78 per cent of the land in the Fens was under the plough, even though much of this had only a few inches of topsoil. Many of the large estates were sold after the First World War.

The Fens at this period was still the major breeding ground for the Shire horse and would remain the last bastion for working horses in the country. The variety of crops grown on fen soils suited the horse, he was kind to these fragile soils, lived off the farm and remained the farmer's best friend. Warns Tinkler, a fenland small-holder on a Huntingdon County Council holding, bought his first tractor in 1951 while John Richardson of Bourne Fen used horses to cart potatoes as late as 1973 on his 1,400-acre estate.

The Second World War brought the last remaining areas of heavy fenland under the plough. Draglines cut new or deepened drains and dykes, and Buckeye draining machines from the USA laid tile drains under the soil for field drainage. The government backed the agricultural industry to the hilt and food had first priority.

A Second World War gun emplacement on the sea bank at Freiston. The land in front of the bunker during the war was out marsh, but it was enclosed in 1972.

J.W.E. Banks Farm Machinery staff on Postland estate in the 1990s.

Lease lend from the USA brought all forms of machinery to the Fens, as well as other parts of the UK. Our fen soils were being tested to their limits in a way they had never been before. Indeed, close cropping of certain crops diseased the soils, which carried on for almost a decade after the war ended and only science allowed us to carry on farming.

The 1970s and '80s were golden days for farming in the Fens. EU grants and subsidies, higher prices, mechanisation, advances in plant breeding and agronomy all enabled us to produce more from our soils. Soil had become a science to produce whatever we wished. The ball was rolling for the supermarkets whose sole aim was to provide cheap, quality food for the public. We as farmers were able to control disease in crops and in the soil, and to carry out most operations irrespective of the weather and soil conditions in a narrow window if required.

The crops produced from our soils have always been governed by the market, wars, famine and, at present, large corporate buyers – those are the rules in a free marketplace. The role of the farmer changes as circumstances demand; in wartime he was asked to produce food at any cost to our soils, as may happen again if nature wishes to interfere in a global economy. At the turn of the twenty-first century, his role was changing from producer to protector of the environment, but change always follows change. We have become masters of our own destiny, slaves to the marketplace, but we are still custodians of our precious fenland soils.

2

The Ultimate Prize

All businesses have resources to fuel the economic wheel of fortune. Many of the resources at their disposal are man-made, by-products, or natural resources. Topsoil is a by-product of nature, conceived over thousands of years through geological transformation. To a farmer his soils are as much a part of his genetic makeup as the DNA in his body. All farmers possess a unique bond between soil and soul, each generation hoping to bequeath their soils, embedded with their sweat and labour, to the next generation to carry on their farming dynasties.

The region covered in this book I will refer to as 'the Fens', although to those who live and farm here the word has many interpretations, with many localities having different identities. Over recent centuries, the Fens were transformed by man from a wilderness of salt marsh and fen into the most prized agricultural land in the UK as a result of the basic need of Britain to produce food to feed its people.

Broadly speaking, the area is divided into 'marsh' and 'fen'. Marsh is land reclaimed from the salt marshes using sea banks to stem the tides, some of which date back to Roman or Saxon times. Fen is land away from the sea that was flooded by upland water run-off. Back in the nineteenth century the black peat fen was the furthest away from the sea bordering the fen edge. The black fens are areas of peat soil, which were originally meres or areas of low-lying land formed by rotting vegetation dating back many thousands of years.

Farmlands between the silts and black fen contain a higher proportion of clay and are referred to by the farmers as 'fen skirt land'. Through here ran the fossilised tidal creeks called roddons (rodhams), or silt hills, as the fenmen call them. These are evidence of past marine intrusions and were formed when water saturated with mostly silt and some clay particles inundated the Fens via the tidal creeks. From the Mesolithic period (c. 6500 BC) and into Roman times (AD 43), these tidal creeks meandered across the land carrying the lighter, paler coloured silts with them. The heavier silt particles were also carried along and ultimately filled the tidal channels. The clay particles were spread over the adjacent areas as the channels overflowed at high tides. The draining of the Fens has left the roddons standing and they are seen as paler coloured ridges in the fields (up to 1m high) because the clays and peats have shrunk due to lack of water. From the air they are visible etchings of our prehistoric rivers and

A roddon running through Thorney Fen. (RS)

The side of a recently cleansed dyke showing the periods of fenland soil history over several thousands of years.

channels depicting the thousands of years of climate change the Fens have experienced.

Many of the fen towns and villages dispersed across the Fens – including Ely, March, Chatteris, Thorney, Crowland, and many more – have developed on gravel islands formed from the last Ice Age, while many smaller settlements, such as Benwick and Prickwillow, were built on roddons. These were also excellent platforms to build the farmhouses and buildings on, being higher ground, more stable and less likely to suffer from settlement in dry seasons. My farmhouse is built on a roddon and there is the occasional farm called Roddon Farm.

This area has been constantly changing for thousands of years, influenced by climate change and the elements. However, the greatest influence has been man himself for without him the Fens may have remained nature's wilderness. It will always remain a challenge to the fenmen to hold their prize so nobly won, that prize being their fenland soils.

The peat fen has suffered more than any other soil types in the Fens, constantly diminishing in area since being drained in the seventeenth and eighteenth centuries and farmed intensively from the nineteenth century to the present day. A book called *The Black Fen*, written by A.K. Ashbury and published in 1958, documents this phenomenon more than any other publication. Ashbury states that he finds it difficult in writing his book to say precisely where the boundaries of the fen peat now lie. His only soil map available of the Fens is one drawn in 1878 by Miller and Skertchly for their book *Geology of the Fenland*.

During the past century many thousands of acres have been reclaimed from the sea, increasing the total area of the Fens. While total area of land has increased, over the same period many areas of black fen have witnessed untold loss of topsoil. A detailed national soil survey was carried out in 1983 by the Cranfield Rural Institute, Silsoe. It illustrates the wide variation of soil types, in some instances over just a few metres. Much of the black fen has lost its organic content over the past century. Through this loss the underlying marine clays and in some place fine silts have appeared and are now mixing with the remaining peat. From the time of the great drainage schemes of the seventeenth century, farmers have endeavoured to improve the soils. The reclaimed peat soils were the most challenging to farm, owing to their high organic content, mineral deficiencies and unstable makeup.

In the nineteenth century, in an attempt to improve soils high in

Cattle on the Whittlesey Lower Washes in 2010, showing the dykes acting as field boundaries. *(RS)*

Claying on black fen showing trenches where peat has been used to fill in the trenches after clay had been dug up and mixed with the peat topsoil. This photograph was taken after the topsoil was removed for road construction, Newborough High Fen, 2010. *(RS)*

organic matter where there was underlying clay several feet down, it was a common practice to carry out 'claying'. This involved digging a trench down to the clay and bringing it to the surface where it was mixed with the peat to improve its structure and cropping potential. The peat from the trench was returned and in some areas the places where this was practiced can still be seen today.

Paring and burning was another method used to improve the organic soils for farming. The early drained peat soils had low Ph levels (lime deficiency) and were called Acid Soils. Lime was not readily available in the Fens but was found in abundance on the limestone lands around nearby Stamford. Liming is a prerequisite for most arable cropping.

Early farmers and landowners had also found that burning the peat soil produces a flux of ammonia, which increases the Ph value making it suitable for arable crops. It also slowed the decomposition of the peat. There were, however, several detrimental effects to this process of improving the soil for cropping: it released copper compounds with the result that it became deficient in copper. Another problem was wastage, which lowered the level of the land, affecting drainage. Fires on the black land were also extremely dangerous. In some areas peat lay several feet below the surface encasing bog oaks, which if ignited could burn for several weeks or even months before they burnt themselves out. Much of the black fen was already experiencing a lowering of land levels below the main drains and since pumps were few and far between, the risk of flooding increased.

For burning to be effective it had to be repeated biannually to maintain high enough Ph levels. It was a common practice, but many landlords restricted their tenants as to how often it could be carried out. However, it still achieved excellent results for entry of wheat or cole-seed and many said 'burning beats the muck cart'.

Efforts were also made to improve the soils on reclaimed land from the marsh. Some sections of the Hay Thompson reclamations in the 1940s had a high percentage of clay in the soil. To improve this imbalance, silt was mined nearby and spread over the land to improve its structure.

The Fens consists of several districts of varying soil types, which give each district an individual

Illustration showing man paring peat topsoil in preparation for burning.

characteristic of its own. In the UK, soils for agricultural purposes are classified in Grades, 1, 2 and 3; Grade 1 being the best agricultural land. Of the total Grade 1 soils in the UK, 50 per cent are in the Fens and these can also be classified into another 3 grades by local land valuers and growers for specialised crops. Almost all the rest of the Fens are Grade 2 soils with only a smattering of Grade 3 around the fen edges.

The very best silts are referred to as 'Toft land' and are of extreme value for growing vegetables. This type of soil is capable of growing three crops of vegetables in an 18-month period, where as the neighbouring field may be a totally different grade of soil. Its ability to hold moisture naturally enables the grower to specialise in crops such as cauliflowers, cabbages, sprouts and broccoli which require constant moisture during the growing period. This grading is reflected in the land's selling price when it comes up for sale where the price can vary as much 50 per cent. Much of the skirt land back in history had been grazing land and the soil beneath the grass had never seen the light of day. Improved draining in the eighteenth and nineteenth centuries transformed much of these areas to arable farming, and what was not put under the plough to begin with was converted to arable during the Second World War. When first cultivated there was a covering of rich, high-organic matter that has now mainly been depleted to around 10 per cent. Similarly, the black fen where organic soils were many feet deep are now no more than a few inches above the estuarine muds. Those soils were very high in organic matter. Almost all soils, silts, skirt and black fen have decreased in organic content over the past 50 years. Previously, with grass leys, farmyard manure and the practice of a wider rotation, the organic content was more stable. I believe we may have reached an equilibrium on organic content on most soils across the Fens at this time.

With the decrease in organic content on silt soils, capping becomes a problem after heavy rains. Areas where the black soils have depleted now blend with the underlying estuarine muds, and consequently are stabilising in their organic content and structure. The soils that have changed little since the major drainage schemes of the seventeenth century are where there is permanent grass, such as the flood washes of Welney and Whittlesey. Cowbit and Crowland Washes are a modern illustration of how soils have lost their organic content in a few decades. They were put under the plough in the 1960s and within four decades have lost much of their covering of organic topsoil. The soils where fruit is grown have changed little, apart from where nutrients and trace elements are concerned. When left to permanent pasture, soils may change in their nutrient values but the soil makeup and structure beneath the grasses does not alter.

The large variations of soil types across the Fens mean that management of these soils can change over short distances. Cultivating machinery for different types of soil can also be totally different on silt, black and heavy clays. Although their soils vary considerably in their makeup, and the management skills required to grow the various crops also varies, the farmer, grower, and grazier have one thing in common: they all use fen soils to produce whatever crops they grow. I remember my father's words when I went to agricultural college, 'Learn the technical and theoretical side of farming, but don't forget how to farm.' Words that at the age of eighteen I did not understand, but I soon did!

The last remaining areas of permanent grass in the Fens are the Hundred Foot Washes and Whittlesey Wash. The grazier never sees his soil only the grass and fodder it produces for his livestock. Unless he reseeds old pastures or sows new leys, he will never come in contact with the earth beneath his feet. He does, however, have to feed his grass due to the livestock removing nutrients from it as well as manage the unwanted weeds in his pastures.

During the post-war years, areas of these washes were ploughed and cropped with arable crops. Being virgin land they were ideal for potatoes and peas and were cropped as such. Arable cropping on flood plains was, however, precarious with the risk of losing the entire crop, with no liability on the drainage boards, since they are flood

Second World War grassland being ploughed up by Caterpillar Twenty-Eight.

plains. By the 1970s food shortages were over in Europe, conservation was high on the government's agenda and the washes returned to their original state, pasture.

Farmers and growers know only too well that what comes out of the soil is only governed by what they put into it, and harvests are determined by their efforts along with nature's. Anyone can own land but the man or woman who lives off the soil has a unique bond with it. It is a part of their heritage, to be treasured and nurtured through their lifetime and passed on to the next generation in as good, or better condition, as they received it. They understand it through all the seasons and conditions as well as its makeup and whether it is deficient in nutrients or trace elements. They know if it is in good heart or not and the consequences of both. Soil has moods and one has to know its moods, to work it or not, to crop it or not, to rest it or not. Some fen soils are more prone to weather factors than others, the silt and black lands are far more forgiving than the skirt and so easier to master. There has to always be an element of respect by the farmer.

The varying soil types across the Fens and marsh have created a breed of agriculturalists whose crops, and methods of producing those crops, is more diverse than anywhere in the UK. These individualists have over time adapted their cropping to befit the many variants of soil types, districts and localities. The list is endless from top fruit, soft fruit, bush fruit and cane fruit to flowers, potatoes, peas, and salad crops in horticulture. In addition, there are cattle, sheep and corn, sugar beet, bulbs, oilseed rape together with mustard and carrots. Crops are then packed, processed, pruned and pared, sliced, peeled, pickled, canned and frozen, and sold often with only the growers' imprint on it. They are indeed masters of their trade at producing products to grace the public's table.

Many of the fenland towns flourished and prospered on a variety of crops and produce grown around them, giving them an individual identity. It may have changed over time but their history does not change. The town of Boston after the demise of its fishing industry became associated with potatoes and vegetables; Wisbech has many

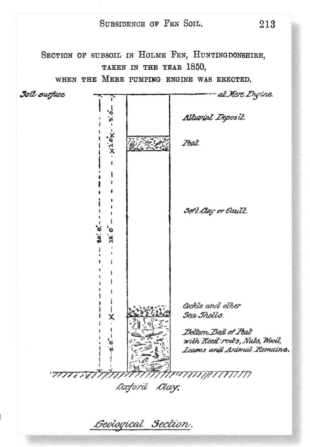

Sections of sub-soil when Holme Fen
was drained in 1850.

types of flower and all types of fruit, while Spalding was and still is synonymous with
bulbs and flowers. King's Lynn, Boston and Wisbech hosted large canning and freezing
plants as did Peterborough, declining over the past two decades. Long Sutton, only a
small fenland town, can boast the largest canning factory in the UK and Whittlesey has
one of the largest frozen chip factories in the UK.

These towns are no longer specialised in individual enterprises, owing to the changes
in demands by the supermarkets, but nevertheless their history lingers with their names.
Today many large packing and processing plants operate in the rural areas of the fen
and marsh where their raw material is sourced, drawing their labour requirements from
far and wide.

The pressure on modern farming methods to produce more from our fen soils is
increasing at an alarming rate. It is being driven by the marketplace whose gearing
involves quality, quantity and timeliness to feed a nation as cheaply as possible. It
has become orchestrated by agronomy rather than husbandry. For our topsoils to
survive the demands we are putting on them, science will have to be their saviour.
Genetically engineered cropping will have to be used to sustain our soils and feed our
nation in the future.

3

Land Ownership

There were and still are several sectors of society that earn an income from the land, either directly or indirectly. Today the majority of landlords in the Fens range from the old aristocracy, the Church, colleges and the Crown Estate to pension funds and local charitable trusts, both family and institutional, many of which have had title to their lands since the Reformation in the sixteenth century.

Farming had been practiced in parts of the Fens since the early Neolithic period. Later settlements through the Iron Age and Roman and Anglo-Saxon periods also lived off subsistence farming. Ownership, as we know it, probably came into being in a simple form with the arrival of the Danes. During the eighth century the Danes began to settle in and around the Fens, during a period of religious dormancy and by the tenth century they were well established and administering their own political system under the Danelaw.

After their political conquest, certain terms associated with the land became commonplace. A hundred was an administrative sub-division of the shire with fiscal, judicial and military functions. A Vill was the word representing an area of land in administration rather than a measured area. A Hide was a standard unit of assessment of tax which would support a household divided into four Virgates, the land-tax levied on that Hide being Geld and the Virgate usually being common land. A Carucate was an area that could be ploughed by a team of eight oxen, and a Bovate was one eighth of a Carucate.

Celtic churches and abbeys were already established in the Fens during the seventh and eighth centuries, some of which suffered during the invasion of the Danes. After the Normans arrived in the eleventh century, the Fens witnessed the great monastic revival over the next 200 years. The Benedictine order established religious monasteries, many of which were on the old Celtic sites of worship. They built abbeys on a grand scale constructed of stone sourced from around the fen edges. The principal ones were at Peterborough, Ely, Ramsey Thorney, Crowland and a nunnery at Chatteris.

With the development of the abbeys we began to witness the beginning of the organised land ownership and management, which would become the origin of the great landlords of the UK. The exact acreage the abbeys claimed title to is not known as these estates were initially defined by boundaries and not acreages, albeit the word 'acre' (Old English 975) cognates with several Nordic words appertaining to 'field'.

Much of the Fens at this period would have been common land, unfenced, and the fenland abbeys did not vie for defined boundaries in the Fens until the thirteenth century. They did, however, put great value and rights on their fisheries such as the various meres. Most of these lay inland, formed by the run-off of highland water into the fen and retained a high water level due to the higher land around the Wash restricting the outflow of water into it.

Major changes in land ownership and management started after the arrival of the Normans in 1066 under William the Conqueror. Norman families established estates here and the abbeys began to take on a more organised feudal system in the countryside. We were fortunate that this period of our history was so well recorded in the Domesday Book, completed in 1086. It recorded the possessions of feudal families. In it we find who owned the land, and how it was managed, as well as a ledger of the animals and ploughs for cultivating soil and the men both bonded and free who lived in the countryside. References in the Domesday Book regarding some fenland settlements include:

Wainfleet; 2 Bovates of land to the Geld   Yaxley; 15 Hides to the Geld
South Kyme; 14 Bovates to the Geld   Sawtry; 3 Hides and 3½ Virgates to the Geld.
Crowland; 3 Carucates to the Geld

The above illustrate the area taxed to the Geld denoting the different grades of soils with Wainfleet being the best, and Sawtry being the poorest soils.

To sustain their extravagant existence and lifestyle and to afford funds for Rome, the Church needed larger estates, which consisted of two parts of their land portfolio. First, their Demesne land was the land required to support the abbey or cathedral. This today would be called 'in hand land' – the produce from the land, lakes, woods, quarries and rivers went directly to the abbey.*

Secondly, holdings within manors were let under 'customary tenure', the freehold being retained by the manor, with varied customs between manors, and the rent not necessarily in cash but in grain, stock and work on the lord's land – often the live and dead stock were leased as well. Land was also let under the term 'Copyhold Tenure'. Copyhold land could be bought and sold, inherited by descendents, left in a will, mortgaged, and settled, just like freehold estates. Many landholdings were held by members of the same family for generations. However, every transfer of land had to go through the lord of the manor and the land was surrendered back to him before the new tenant was admitted. The lord of the manor had the right to take fees from new tenants, and to receive a payment called a 'heriot' on the death of one of his old tenants. The official record of the transfer of copyhold land was written up in the manorial court rolls. In addition, the steward of the manor wrote out an official copy of the court roll entry, which was kept (held) by the tenant as their proof of title. This is where the term 'copyhold' comes from.

Other parts of the Church estates were made up of manors and granges, each with their own Sokeland, this being land appurtenant to a manor or grange. The 'freeholds' still belonged to the abbeys but were managed by the lord of each manor or grange – in today's terms they could be described as the tenants of the abbeys.

Warland was land held by peasants and tenants as opposed to their lord with obligations, especially for the payment of Geld, or services. A Berewick was also used in the Domesday Book, being an outlying estate often devoted to some specialised function. An example of a manor in the Fens in 1222 was Tydd St Giles. The manor consisted of 1,071 acres – the 'Demense' land was for the exclusive use of the lord of the manor, the remaining 600 acres let to free-tenants and copyholders, copyholders being the largest holders of land.

Robert, son of Walter, held 256 acres, Stephen de Marisco held 112 acres and a widow named Innocencia held 2½ Virgates (80 acres). Affairs of the manor were managed by the manor court or court baron which was held annually. Much of the land around Tydd St Giles was either marshland or land liable to be flooded by the incoming tide or even waters from the upland area coming through the Fens. Today the parish covers five times the land mentioned in 1222, due to reclamation. The total rent of the land in 1222 was £4 3s 7d, 4 score hens, 600 and 4 score eggs besides the services of the tenants.

As well as the fen and fen edge, the marsh was also a bountiful source of revenue from salt-making, and the harvests of fish, fowl and samphire claimed by the abbeys and some Norman families. There are several mentions of these families in the Domesday Book. Kolgrimr had land in Dowsby and salterns in Bicker while Guy de Craon had land at Rippingale, Dowsby and many more areas on the fen edge. His manors included Sokeland and fertile areas along the marsh, such as, Kirton, Butterwick, Wrangle and Fishtoft, as well as a salt pan in Bicker. With so many abbeys in and around the Fens, their ownership of pasture, woods, tillage, fisheries and salt pans was large. Their assets were not confined to the abbeys' boundaries but spread far and wide.

---

* Demense land 'in lordship' whose produce is devoted to the lord rather than his tenants. When applied to an abbey, the produce was of the abbot.

Many abbeys from various other parts of the country also held interests in the Fens, as did the king and other families. It is evident through the extensive ownership of land where salt-making was carried out how important this industry was, especially in Lincolnshire. The Bishop of Durham, Archbishop of York, Bishop of Lincoln and the Abbot of Ramsey each held rights to salt pans, while the Abbot of Peterborough held sixteen salt pans in Bicker. Bicker had the largest concentration of salt pans in the Fens. It was because of the abbeys owning the salt pans in Bicker that it was not enclosed to the tides until the seventeenth century. We presume that salt must have been more lucrative than fertile farmland. The same must have applied at Holbeach Hurn, which at that time was on the seaward side of the Roman Bank.

By the twelfth century other religious houses had joined the Benedictines in the Fens, including the Cistercian, Gilbertine and Augustinian Orders. All the abbeys in the Fens were spread either around or in the Fens itself, except for on the east section above Ely on the Norfolk-Suffolk borders, where there are none. For four centuries this area remained the domain of the monasteries and Norman families with little drainage of note and probably few changes of land ownership. The Reformation in the sixteenth century changed this dramatically, with large areas becoming the property of the monarchy, from Henry VIII's reign through to James I's. There was also a transfer of demesne land from the great fenland abbeys to the courtiers of the king. This was the beginning of the system of absentee landlords which would remain until the late nineteenth and early twentieth centuries.

Most of the country was pastoral farming which requires a small labour force. With the reduced labour force available to work on arable land, and the lack of demand for grain foods, large areas switched from arable to pastoral farming or grew nothing at all. The Corn Laws of 1360 were still in place protecting the incomes of farmers and large country estates where arable farming was practiced.

By the seventeenth century, the population was increasing and so began a great push for England to be self-sufficient in food, and the Fens would become the 'Klondike' of agriculture. Up until the fifteenth century, land was let to tenants 'at will or pleasure'. Customary leases were mostly short-term, but some were longer where tenants could sub-let their land, and copyhold. The tenure of land began to change in the fifteenth century with copyhold tenure, 'by copy of the court roll according to the custom of the manor,' resulting in longer fixed terms of letting, even lifetime, became more commonplace.

During the late eighteenth and early nineteenth centuries it was still common in the Cambridgeshire Fens on newly enclosed land to let land on restrictive covenants, rather than leases. During the nineteenth century large tracts of the Fens and marsh were still owned by aristocratic and wealthy families from outside this area. The colleges, Church, Crown and Trusts, such as Guys and other hospitals, were also important landlords. Most of the people farming the soil were tenant farmers with no real security of tenure. Farmers who did own land must have been few and far between, especially in the Fens.

Today, the Childers Trust owns 1,700 acres of land at Eldernell near Whittlesey and is probably one of the oldest private land-owning families in the Fens. John Walbanke-Childers of London was MP for Cambridge and died in 1862 leaving two daughters. His mother's father was Sir Sampson Gideon I and last Lord Eardley whose wealth came from Sampson Gideon 1699-1762, the son of a naturalised Portuguese Jew who was a West India Merchant in the City of London. He amassed a fortune, even helping the British government's finances in times of great need, and bought landed estates in Lincolnshire and Buckinghamshire. One such estate was in Borough Fen and consisted of 3,000 acres, including the duck decoy, land at Spalding (lord of the manor) and other parts of the Fens.

The period from the agricultural depression from the 1870s through to the First World War would bring about the birth of the owner-occupier in the Fens. In 1909 only 13 per cent of the land in England was in the hands of owner-occupiers. Due to the depressed state of British agriculture many estates were put on the market, with much of the land being bought by the tenants.

The largest fenland estate ever to be sold was the Thorney Estate, an area of some 20,000 acres, which was offered to the tenants in 1910 (see section on Bedford Estate). Some estates changed ownership from one landlord to another, without tenants being offered their farms. It was the birth of a new breed of farmers in the Fens. There was no food rationing during the First World War and their financial salvation would turn out to be the war, when food was vital for the country, and prices of their commodities escalated allowing them to repay the loans taken out to buy their farms.

After the First World War land prices rose on the back of farming being profitable once again. This was an opportunity for those remaining landlords to cash in on the higher land prices. Not all wished to sell, though many were forced to as a result of family members being killed in the wa, meaning death duties had to be found. Others, remembering the long depression in the late nineteenth century and pre-war years, when revenue from their estates was either nil or at best nominal, also took the decision to sell. Some families accumulated large areas of land in the fens and marsh, while others were content to purchase just their own farm on which they had been tenants. Many tenants also bought on 'back-to-back' deals, where they bought the farm at a tenanted price and sold it to another landlord, retaining their tenancy.

Charities for the poor have, since medieval times, been a great part of Fen village and town culture, almost every one still having several today. Before the nineteenth century poverty was rife in the Fens hence almost all these charities consisted of land for the poor, or land let to provide monies for them. Many of these benefactors were described in early nineteenth-century ledgers as 'yeomen', being someone who cultivates his own land. The term 'Yeoman Farmer', however, took on another meaning later in the century, describing them as a highly regarded group of landowners second only to the landed gentry, which the Fens were beginning to witness.

Potatoes were established as the staple diet of the urban communities during the late nineteenth century, as were fish and chips as the new fast food of that era. Fast rail links with the outside world and virgin land to grow the crop would set the stage for the birth of 'Potato Barons' in the Fens. This was probably the only sector of agriculture which did not suffer in the depression between 1870 and the First World War.

After the First World War, farming went into a depression again, and on the back of it the land market started to decline in 1921. Landlords were eager for tenants, and in some cases offered new tenants rent-free farms for the first year or two to allow them to get established. My family took on two rented farms in the period 1927–35 rent-free for the first two years. In 'The Land Market 1880–1925' in the *Agricultural History Review*, Michael Thompson asserts that 'owner occupation increased from 10.9 per cent of the cultivated area of England and Wales in 1914 to 38 per cent in 1927, an increase of 2.9 million acres.'

Following the pattern of pre-First World War years, farm incomes increased in the years leading up to the Second World War, and during and after the war, with little land sold during the war years. Land prices rose after the war, as they had done after the first conflict. Farmers thought that farming would follow the trend of the depression years of the late 1920s and early 1930s. This time, however, it was the sellers driving the market. Instead of landlords selling as they had done after the First World War, it was tenants who sold their farms on 'sale and lease back' (SALB) deals to institutional and private investors. This trend carried on during the 1940s and through the 1950s.

MICHAEL BROWN, W.H. BROWN (W.H.B.) OF SLEAFORD

From the 1950s through to the 1980s, W.H. Brown was at the forefront of the sale and lease back market throughout the country, the Fens especially. This form of purchase is where a buyer purchases the freehold land from a farmer, who then remains in the farm as the landlord's tenant. SALB deals were not new in the Fens – Trevarthoe House Farm in Holbeach Marsh was bought on SALB from the Tinsley family in 1858, which was probably one of the earliest times this mechanism is recorded. W.H. Brown's first sale and lease back was a 300-acre farm at Old Leake, Boston, in the 1950s in conjunction with Bidwells of Cambridge to one of the Cambridge colleges. Michael's father did the deal and the price paid was £41,000 on a full repairing lease with the rent being £5 per acre. Other notable buyers in the Boston area in the mid-1950s were the Wills tobacco family and the Merchant Adventures.

W.H.B. sold Cranmore Lodge, Deeping St James, a farm of 437 acres on sale and lease back for £141 per acre to a private buyer. Other SALB deals transacted by them during the 1960s were: 1962, 2,000 acres; 1963, 4,912 acres and 1964, 4,786 acres. The words 'they don't make land any more', a cliché used by a few land agents to tempt the City investor out of his office to get mud on his shoes, would rock the land market like never before. This was a virus that was on the tips of the land agents' tongues, and would spread across the City institutions like a plague. In the late 1970s institutions, pension funds, colleges and Crown estates rushed into the market for land as it became a fashionable part of a portfolio to hold. Entry into the EU encouraged farmers to re-equip their machinery and buildings to cater for the growth in supermarket demands. Growers' co-operatives funded by AMDEC grants from the EU sprang up like weeds. The land market went mad for SALB deals, not only in the Fens but across the whole country.

In the first months of 1972, farmland prices rose 250 per cent, reaching record highs in 1977, which held until 1978. Most of these sales were SALB deals with the margin between freehold prices and SALB prices narrow, so desperate were institutions to own land and farmers to retain tenancies. By 1978 financial institutions owned 11 per cent of freehold farming and forestry land in the UK. Within a few years, thousands of acres in the fens were sold back to the previous owners at prices well below what they had sold it for. With my father, I attended a sale of a farm in Thorney Fen that we were interested in. The bidding kept going up and up between agents acting for two pension funds and was eventually knocked down at a very high price. My father turned to me and said, 'I wonder if these City boys know whether they are buying acres or hectares.' The buyer sold the farm some years later for half the purchase price. It was not only the City speculators who came to the Fens. The Duchy of Cornwall's agents could not resist a piece of the Fens and bought farms in Thorney and Bourne Fen – these have also since been sold on.

This 'land rush' was, in hindsight, a great benefit to fenland agriculture in several ways. Those farmers who sold land and ceased farming were left with cash reserves to either reinvest or retire with. The ones who sold on SALB deals and bought their farms back at lower prices either cleared their debts, or had surplus cash to expand their farming enterprise. Much of the land bought by the pension funds and institutional investors was farmed by large farming companies created during this period. This was the birth of 'contract farming' agreements, a major change in farming relations between landlord and tenant, and some of these companies remain today. Many investors who had purchased farmland made further investments in redraining the fields, erecting new buildings and carrying out tree-planting. They came in a hurry and left in a flurry, but left us a legacy.

Farmland prices in the twentieth century seldom related to farming incomes but more to investor fever, tax avoidance through government legislation, or roll-over money from land sales for development to avoid taxes. In 1976 the Labour government brought in the 'Succession of Agricultural Tenancies' which enabled a farm tenancy to pass to the tenant's children for their lifetime, conditions being that they had been actively involved in the farm, and the tenancy was a vital part of their income. It was supposed to safeguard the next generation of farmers but in fact it deterred landlords with vacant land from reletting, and prevented new entrants into farming.

Another scam to avoid tax by buying farmland was around in the 1970s. If farmland was passed on to someone in the family, after a period of seven years it became free of Estate Duty. When Capital Transfer Tax was introduced to replace Estate Duty, this seven-year period was abolished. In addition, if someone with wealth and readily available cash was about to die they could buy themselves a place in heaven. The government brought in a 'Death Bed Concession' of 45 per cent Death Duty relief with no strings. This relief on farmland, with no restrictions on length of time it had been owned or farmed, became a bonanza. With land prices so high it did not need many acres to be available for this to be a lucrative loophole. Farms were bought just hours before someone died and were sold back again within days. The vendor, the agent and the purchaser divided the spoils of the 45 per cent relief – this was abolished in 1978.

The land market began to slide from November 1979 to March 1981 perhaps by 15–20 per cent, but then recovered by the summer of 1981. In the 1980s costs caught up dramatically, with the quadrupling of oil prices and rising inflation causing farm borrowing to increase by 36 per cent to £1,500m. However, best silt remained in demand still selling for more than £10,000 an acre in 1983.

There are without doubt many more owner-occupiers in the Fens now than in the past, but substantial landlords still remain with their land being let to tenants on Farm Business Tenancies (FBT) or Contract Farming Agreements. We are going through a period of decreasing land sales, although sometimes estates change ownership without the tenants knowing. Many owner-occupiers have ceased farming and contract farm their land to larger farming organisations, not wishing to sell their farms thereby being eligible for Estate Duty relief. This is probably the main factor why land sales have decreased in the past decade, coupled with a more stable farming industry.

The short FBT have been good for those farmers renting the land, and those who have used them to expand the structure of their fixed costs by farming more acres. It remains to be seen in the long-term if they will be beneficial for the soil's well being. If farming incomes decline, many may have to exploit it more to survive. The introduction of FBTs was to make the letting of land more flexible and encourage new entrants into farming, which has not materialised; indeed, quite the reverse has happened. If any grade of land in the Fens is put on the market it realises the highest prices per acre we have ever witnessed, and they still 'don't make it any more'.

The Crown Estate as a Landowner

The largest estate in the Fens is owned by the Crown Estate Commissioners, although exact acreage of fenland is not easy to quantify. Parts of their estates are on the fen edge and since it is difficult to define where the Fen ends and high land starts, exact acreages cannot be given. However, an approximate acreage is around 62,500 acres (25,000 hectares), thereby making their fenland estates 8 per cent of the entire Fens. Their total portfolio of agricultural land, including forests, is 360,000 acres (146,000 hectares) in the UK.

The Crown Lands date back to the Norman Conquest when most of the land was owned by the monarch himself. After the Norman Conquest, all the land belonged to William the Conqueror 'in right of the Crown' because he was king. The Sovereign's estates had always been used as a means of raising revenue, and over time large areas were granted to nobles. The estates fluctuated in size and value but by 1760, when George III acceded to the throne, the asset had been reduced to a small area producing little income – revenue which George III needed to fulfil the Sovereign's fiscal responsibilities to the nation. By that time taxes had become the prime source of revenue for the United Kingdom and parliament administered the country, so an agreement was reached that the Crown Lands would be managed on behalf of the government and the surplus revenue would go to the treasury. In return, the king would receive a fixed annual payment – today this is known as the Civil List. This agreement has, at the beginning of each reign, been repeated by every succeeding Sovereign. In 1955 the estate changed its title from 'Crown Lands' to the 'Crown Estate'.

The Crown's fenland estate has been amassed in size over many centuries through claim of ownership, purchases of freehold land and purchases from owners who remained as tenants (sale and lease back). Land reclaimed from the sea together with the Enclosure Acts have also swelled the acres of the Crown in the marsh areas. The Crown Estate, like the colleges, Church Glebe and Church Commissioners who own land, have always been landlords, never farming their land themselves (in hand). I did find, however, that in 1904/5 they did farm a 917-acre farm for 2 years which they may not have been able to find a tenant for, due to the depression years. In 1906 it was divided into 83 small-holdings, 680 acres with 12 tenants and 237 acres with 71 allotment holders.

Many of their tenants are long-standing and highly respected by the Crown's land agents. The Crown Estate is comprised of small and large blocks of land spread across the entire Fens consisting of fen and marsh together with a diverse range of soil types. With this range of soil types the cropping on their farms is varied. It covers the entire spectrum of crops, ranging from the combinable cropping on heavier soils, to the light silts growing salads, roots and vegetables and the fen peat soils growing roots, onions and salad crops. Its tenants range from the very large to small-holdings on the silts soils.

The main blocks belonging to the Crown Estate are at Billingborough, Holmewood, King's Lynn, Wingland and includes land around Holbeach, Spalding, Whaplode and Moulton districts.

HOLMEWOOD ESTATE, CONNINGTON

This estate is made up almost entirely of peat soils with a subsoil of marine estuarine muds. Bog oaks are still being found near the surface. Much of the estate is on the site of the Whittlesey Mere which was drained in 1851–2.

Bog oaks being cleared from a field in Holme Fen in 2003 after recently appearing. (RS)

The estate was owned by Squire Wells who set about the reclamation to fund the work. He mortgaged the estate to Lord De Ramsey who foreclosed and sold the estate to Squire Feilden. Feilden donated a hospital ship in the First World War and a wing of Spitfires in the Second. He was also a bachelor and left the estate on his death to a hospital. The hospital declined the bequest and so the estate was put on the market. The estate was bought by the Commissioners of Crown Lands in 1947 from the executors of J.A. Feilden for £160,000. The estate was subject to considerable dilapidation at the time of purchase. Following the purchase, the mansion with 75 acres was sold to J.J. Wells, a descendant of the former owner of the estate, having previously been occupied by the Sugar Beet Corporation who used it to store records. A further 640 acres was sold to the Nature Conservancy – mainly woodland.

Today the land is farmed very intensively with root crops and onions. This type of land was not readily sought after for farming until advances were made in agronomy after the Second World War, and the introduction of irrigation together with modern machinery. Like most land furthest away from the Wash, by nature being the lowest in the Fens, improved drains, larger pumps and more pumping stations added to its success as farming land. However, loss of topsoil through shrinkage over time has been a concern together with bog oaks appearing on the surface hindering agricultural machinery.

The estate acreage has changed over the time the Crown has owned it and today stands at 1,264 acres. In recent years 3,728 acres have been sold to accommodate the Great Fen Project along with 700 acres they have options on.

KING'S LYNN

The Norfolk Estuary Company was set up in 1846 to carry out reclamation around the Great Ouse estuary (see chapter 5). The Crown Estate arose out of the purchase of the Norfolk Estuary Company's (NEC) property interests in 1964. Following this, the reclamation of 700 acres at King's Lynn proceeded in 1966. A reclaimed area in front of Sandringham was then sold on to Her Majesty personally. Part of the estate sold to the Queen bordered a bombing range in the Wash. The purchase was driven by Lord Perth, then chairman of the Crown Estate, who took a great interest, which included an overflight by a helicopter in which the aircraft strayed onto the bombing range which was then active.

In the mid-1980s further proposals for reclamation came to a halt, partly because of a planning moratorium imposed by Norfolk County Council following a similar move by Lincolnshire County Council and by the simple fact that the sums simply did not add up (land values versus costs). This primarily affected the Wingland foreshore. Today the estate stands at 18,600 acres of which 1,100 acres is salt marsh.

BILLINGBOROUGH

The bulk of this estate was purchased from the Earl Fortescue when he sold his Sempringham Estate in 1855 to the Crown Estates. The total acreage of this estate today is 14,000 acres with around 60 per cent being fen and fen edge, the remainder being high land, out of the fen. The farms are mostly large being 500 acres or more, the smaller ones being in the Fens.

WINGLAND

The present estate covers an area of 9,240 acres. In 1870 the estate was recorded as totalling 3,107 acres, including land to be allotted to the Crown and purchases from the Nene Commissioners. Lighthouse Farm was embanked between 1910 and 1917 by the Ministry of Agriculture, Fisheries and Food (MAFF) to form part of their Sutton Bridge Estate. In 1926 it was sold to the Commissioners of Crown Lands in 1926 along with Nene Lodge Farm. The tenants of Lighthouse Farm (Bass & Wright, 1929) apparently serviced the holding using a light railway system with temporary sidings to clamps where necessary.

Further to government policy in 1907, small-holdings were established on the Wingland Estate by breaking up existing larger holdings. A series of leases from 1909, 1910 and 1914 to the Holland County Council were made, and the council in turn sub-let them to occupational small-holders. Other areas were let to MAFF and again sub-let to small-holders. Land was leased to the Sutton Bridge Urban District Council, which also sub-let to the Lincolnshire and Norfolk Small-holders' Association from October 1907. Other land was added and leased for twenty-one years from 1913, expiring in 1934. There were around forty small-holdings on this land. Jephson Hall was taken over in lieu of dilapidations when the lease fell in.

The 1925 enclosure was the first to be undertaken by mechanical means with the use of adapted 14-ton steam ploughing engines. There were no further reclamations until 1951. Purchases on sale and lease back also proceeded – eg. Banklands 1964 and Greenlands 1970. Three further blocks were reclaimed in 1951, 1955 and 1974.

WHAPLODE, MOULTON, SPALDING

These three parts of the estates are the most fragmented of all the fenland estates, covering 6,780 acres. The Crown's ownership of these lands dates back the longest of the entire fenland estates. Their titles relate to purchases over many centuries, enclosures and ancient rights. In AD 868 the ownership of lands in and around Spalding and Whaplode were confirmed by Burgred, King of Mercia, to the Abbey of Crowland, with whom they seem to have resided through the Conquest right up to the Dissolution.

The Crown's interest at Whaplode originates in the ownership of the Manor of Whaplode Abbotts (Whaplode Abbitas) and the Manor of Moulton. These three estates came to the Crown on the Dissolution of the Monasteries. The Manor of Moulton was held by the Prior of Spalding and leased to John de Moulton for £7 9s 4d. On his death the manor was divided between his three daughters, Joan Fitzwilliam (the Fitzwilliam fee), Elizabeth Harrington (the Harrington fee) and Margaret de Lucie (Moulton Dominorum). De Lucie had no heirs and this element – Moulton Dominorum – was divided between Moulton Harrington and Moulton Fitzwilliam. The Harrington fee manor with the moiety of Moulton Dominorum fell into the ownership of the Duke of Suffolk and on attainder in about 1450 was forfeit to the Crown.

In 1793 land was added to this estate by an act of parliament passed for the embanking of lands in Spalding, Moulton, Whaplode and Gedney where there is a long history of embankment from Roman times onwards. According to 'ancient documents', the reclaimed land was divided into six portions, the portions being divided to the Lord of Manor of Spalding (Lord Eardley) (one portion), Moulton Harrington (two portions), Moulton Fitzwalter (one portion) and Moulton Dominorum (two portions). This was the same year the South Holland Main Drain was cut from Cowbit to Peters Point, Sutton Bridge.

There was extensive reclamation carried out along the entire Wash coastline from Fossdyke to Sutton Bridge during the eighteenth century. By the early nineteenth century the Crown's interests were in lease to Lord Eardley, who submitted a claim to the Whaplode land as lessee on the passing of the Holbeach and Whaplode Commons Act of 1812 and other enclosure acts from which flowed a series of awards – the Spalding and Moulton Award 1811, Holbeach and Whaplode Award 1819 and Deeping Spalding Cowbit Award 1819 as well as Moulton Common Enclosures. This effectively resolved the Crown's interests in common into wholly-owned parcels of land.

The enclosure awards were followed by a series of purchases. In this the estate followed the classic pattern of the post-1811 reorganisations of promoting enclosure acts then, where appropriate, using the allotted lands as a nucleus on which to build an estate by additional purchase.

On 24 April 1857 the 684-acre Holbeach estate was purchased from William Butcher for £32,000 (£47 per acre) and let to a tenant the rent being £1,200 per annum (£2 per acre) giving the Crown a yield of almost 4 per cent on their investment. Trevarthoe House Farm was bought on a sale and leaseback basis from the Tinsley family in 1858 – which was probably one of the earliest times this mechanism is recorded as being used by the Crown. Quite apart from the sale and lease back, it was an estate of tenant innovation; the tenant, Mr H.C. Tinsley was using portable 'wirelesses' to communicate to his men in 1952. The estate made further purchases of land including Holme Farm in 1954, St James's in 1958, Fir Tree Farm in 1959, and Vicarage Farm in 1971. The land at Friskney was purchased in 1989 and 1992.

This part of the Crown Estate is not 'ring fenced' and has never been part of a larger estate and like Wingland, was at the forefront of the small-holdings movement and even today consists of mainly smaller holdings.

The Crown Hall Farm Moulton Eaugate, part of the Crown's Moulton Estate, is the largest single holding in this part of the Fens and has been in the tenancy of the Clarke family since 1856. Many of the smaller holdings in the Fens area of the estate have been amalgamated as they became vacant to form larger, more sustainable holdings for the tenants. There are some large tenant holdings at Wingland, King's Lynn and Moulton, consisting of units more than 1,000 acres, some even larger.

4

Allotments and Small-holdings

As one drives across the Fens with its vast expanse of fertile soils it conjures up the perfect environment for large-scale farming with mammoth hi-tech farm equipment. This is the case in many areas of the Fens and marsh, but intermingled among those large agri-businesses are small-holdings. They are not as small as they were intended to be 130 years ago when the idea of small-holdings and allotments was conceived, but they are still there.

The National Estate of County Council Holdings in 2007/8 was around 121,410ha (300,000 acres). The county councils of Norfolk, Cambridgeshire and Lincolnshire have cumulative land portfolios of approximately 23 per cent of the total of the whole country. These three counties had a combined holding of 28,595ha (70,657 acres) at 2008/9. The divide between Fen and upland soils is not a definite line, making it somewhat difficult to give a precise figure of fenland involved. If detailed surveys were carried out it would be fair to say that between 60 per cent and 65 per cent of the entire holding could be classed as fen soils. All three county councils have, since the peak of their land holdings, sold off land, mainly out of the Fens. Today their fenland accounts for more than 15 per cent of the entire national county council land holdings.

The history of the county council holdings does not go as far back as the Crown, the colleges, the Church, or even some of the private landlords in land-owning terms. In fact, they are relative newcomers as agricultural landlords.

Allotments were part of our food-providing structure even before the Norman Conquest. More recently, acts in the nineteenth century to enclose land were depriving the working class of a source of food and endangered their very existence. The General Enclosure Act of 1845 and later amendments, however, attempted to provide better protection for these people. The act required that the commissioners should make provision for the landless poor in the form of 'field gardens' limited to a quarter of an acre. This was really the beginning of allotments as we know them today. In the Allotments Extension Act of 1882 an allotment is defined as 'a small piece of land let to a person to be cultivated by him as an aid to his sustenance, but not in substitution for his labour for wages.' Most allotments were owned by the larger land-owning families.

The public ownership of land began during the depths of the 'Great Agricultural Depression' between 1870 and the First World War. The rural economy was at its lowest ebb in history, both for the farmer and the farm worker. The government realised that land was extremely cheap to buy and by creating small-holdings it would enable new entrants into this depressed industry, and create a new breed of agriculturists.

However, the Local Government Act of 1894 introduced a new scheme, whereby parish councils could take over the existing allotments in their parish. Larger holdings of land, termed 'small-holdings', were seen as an advancement of allotments and would be a rung on the ladder to enter farming. Many larger farms were in a dire state and this was seen as a solution for the poor and unemployed.

Their initial solution was the Small-holdings Act of 1892, empowering county councils to purchase farmland and create small-holdings. Most of the farmland was owned by the landed gentry and let to tenant farmers, large and small, when tenants at that time had no security of tenure.

Up until the First World War these tenanted farms had not been profitable, so the feeling among the humanitarian politicians was to look at ways of changing the face of the countryside. The yeoman farmers had also suffered, with many dropping by the wayside during the depression. Many farmers either changed from arable farming to livestock, requiring less labour, coining the phrase 'dog and stick farming', or they sold their land if a buyer could be found.

The late nineteenth to the early twentieth century was the period of mass imports of food from around the world, fostering a feeling among the politicians that British agriculture was not needed when cheap food could be brought in. The government's idea was to give people the option to farm in their own right, 'a ladder of opportunity', where small-holdings provided 'starter units' from which tenants could progress on to larger farms either on the CC estates or in the private sector. Holdings could be of any size, from the small 1-acre allotment near a village or town, to larger self-contained holdings of 40 acres with their own house and buildings. New entrants could start farming on an allotment and graduate over time to become 'a 40-acre farmer', a term which in time was to signify a position of status.

The first of the Fenland CCs to respond to this Act was Holland County Council in Lincolnshire. In 1893 it leased a 24-acre allotment in Holland Fen from the Revd Mr Turk for £2 per acre. It was on a 10-year lease and was renewed in 1903, 1913 and 1923. The council's first purchase was also in 1893, when it purchased a small-holding containing 88 acres of land from Mr Welby for £3,750 (£42 per acre). This was called Pelhams Lands.

The following year, in 1894, 48 acres were purchased in Freiston followed in 1895 by the purchase of 31 acres called Chapel Hill Farm from J.H.G. Basill for £1,675 18s (£54 per acre). In 1897, 46 acres were also purchased in the parish of Tydd St Mary with no mention of the vendor. Their land portfolio did not then change, from 231 acres owned, 24 acres leased, until 1908.

The beginning of the twentieth century witnessed not only the depression in the countryside but also rising unemployment in the cities and towns across the whole country. Under the Liberal government of Herbert Asquith, the Small-holdings and Allotments Act of 1908 made it a duty for county councils to establish small-holdings and gave them the power to compulsorily purchase land. It was hoped that this act would stem some of the unemployment and unrest, and raise the standard of life for many of the population.

It was not only councils who were providing land for small-holders; some private estates in the Fens were doing so before the councils mentioned later in this chapter. It set the scene for more land purchases by the main fenland county councils in Lincolnshire, Cambridgeshire and Norfolk, as well as the rest of the country. Agricultural land was readily available and many acres were purchased by all three counties after this act up to the First World War. Holland (Lincs) increased its holding by 749 acres and by the war had 7,499 acres in the Fens. Little land was purchased during the war but buying began in earnest again after the conflict when Holland (Lincs) made two more purchases totalling 4,266 acres from three vendors. This brought its land portfolio to 12,932 acres returning a rental income of £37,121, from 1,076 holdings, making the average holding/allotment 12 acres. The total number of tenants was 1,101 made up of 844 non-resident tenants and 257 resident tenants.

The post-war years witnessed many family estates being sold due to death duties as a result of the war and by some of the larger land-owning families, many of whom were pleased to rid themselves of the burden of land and tenants alike. These estates had returned little, if any income, during the depression years and with land prices before the war at the lowest for many years, many estates were unsaleable.

The market did however recover after the war when many owners saw this as a time to depart from landowning, knowing the days of the landed gentry were numbered. There was a great push to break up these larger holdings into smaller units for the good of the countryside and its people. The war had changed the attitude towards British agriculture, when supplies were not always available from abroad.

Norfolk CC purchased its first small farm of 91 acres at Chapel Farm, Nordelph, near Downham Market, in 1904. This area had been developed for small-holdings previously through the Norfolk Small-holdings Association by Sir Richard Winfrey. (see page 36.) as far back as 1899. As for many other county councils, the Small-holdings Act of 1908 was the impetus for Norfolk CC to buy land for its land portfolios.

From the passing of the Small-holders and Allotments Act in 1908 to the outbreak of the First World War in 1914 the council acquired 14,000 acres. No purchases were made during the conflict but 5,000 acres were added afterwards, when buying recommenced and the slump in land prices came. By 1951 their portfolio amounted to 31,000 acres, of which about 11,000 acres lay in the Fens. This estate at its peak was equivalent to 2 per cent of the total area of the county itself – today it is approximately 17,300 acres of which 9,171 acres is fenland.

Cambridgeshire CC's first purchase was in 1892 with further purchases in 1908, 1918, 1930 and remained buyers for land up to 1973. No purchases were made during either of the world wars although between the two conflicts the number of tenants grew to 3,000.

In 1984/5 the estate consisted of 44,376 acres of Grade 1, 2 and 3 soils types made up of 964 holdings, which averaged 46 acres per holding. Their largest block of fenland was near Warboys, being 4,500 acres of Grade 1 soils. In 1993 the number of tenants had declined to 600, with farms ranging from 2 acres to 400 acres in size. Their policy, like other CCs, was to make holdings larger and more viable for tenants. Sales of building or development land have taken place in the past, and this is still their policy if favourable opportunities arise. In 2009 the Cambridgeshire CC still had 33,965 acres farmed by 257 tenants.

THE CROWN ESTATES

The Crown Estates was one of those on which small-holdings were established by breaking up existing larger holdings, to further government policy. Further to the Small Holding and Allotment Act in 1907, small-holdings were established on the Wingland Estate by breaking up existing larger holdings after which the Crown Estate also made a series of leases in 1909/10/14 to the Holland County Council who in turn sub-let them to occupational small-holders. These leases have been rolled forward ever since. Other areas were let to the Ministry of Agriculture Fisheries and Food and again sub-let to small-holders. Land was leased to Sutton Bridge Urban District Council and the Lincolnshire and Norfolk Small-holder Association from October 1907. There were around forty small-holdings on this land. Jephson Hall at Wingland was taken over in lieu of dilapidations when the lease fell in. This policy was opposed – too vigorously – by Cluttons, the then receivers, with the result that they were replaced as the Crown Estate agents by Carter Jonas, who were the agents for Lord Carrington, President of the Board of Agriculture. Harry Carter observed, 'the ironic thing was that in agricultural estate management terms they were 100 per cent right and so are to be respected for having taken this line.'

The Crown's Spalding Estate was also involved in the small-holdings movement. Moulton Parish Council took a lease of twenty-one years from 1907 with two other blocks in 1913 then rolled forward. This was rather unusual in that county councils were generally the small-holding authority. This point survived several queries including one by the Ministry of Health (what their interest was has not been researched) in 1950. It was, however, assumed that the parish had acted lawfully. MAFF also managed other areas under the Small-holdings Colonies Acts of 1916 and 1918. In 1912 Moulton Marsh Farm was leased to Holland County Council and divided into small-holdings. MAFF also managed other areas under the acts, and these were surrendered to the Crown Estate in the 1980s and the tenants became direct tenants of the Crown Estate. Since this time a policy of amalgamation to produce viable holdings in the moving agricultural climate has been followed.

LORD CARRINGTON

A landowner at the forefront of the development of small-holdings during this period was Lord Carrington, owner of land in the Fens. In 1887 His Lordship freed some of his land near Spalding to the Spalding Allotments Club and in 1894 another 85 acres to the Holland CC first small-holdings. In 1902 the estate let 650 acres in Deeping Fen to the South Lincolnshire Small-holders Association on a twenty-one-year lease. They sub-let the land to 170 tenants where previously it had been farmed by five or six. 115 of these tenants had allotments of 1 to 2 acres. The association charged the tenants 40s per acre, and they in turn paid Lord Carrington 33s per acre. Landlords such as Carrington now had a guaranteed rent for their land, something they had not had for the previous forty years during the Great Depression. The total amount of rent paid by the 170 tenants was £1,300 with the exception of a trivial amount of £1 13s in arrears.

Cottages and farm buildings were built on the holdings for those tenants who were resident, roads put down and many other services provided – this was an added investment for the landlord. This capital for cottages was borrowed from the Lands Investment Society at 4 per cent interest over a forty-year period, tenants being charged £2 10s an acre. It was prized to be able to obtain a small-holding, even without a cottage. Many of the non-resident tenants travelled up to 8 miles to work their holding, even crossing Cowbit Wash during the winter every day to feed their livestock.

According to *Country Life* magazine in 1906, Queen Victoria honoured Lord Carrington with an Earldom, the Earl of Lincolnshire, in 1895, and King George V elevated him to Marquess of Lincolnshire in 1912. His only son was killed in 1915 during the First World War, hence both titles became extinct. His brother succeeded him in the barony and the family still have land in Deeping Fen today.

THE GREAT REFORMERS

Richard Winfrey was born to fenland parents on 5 August 1858. In *White's History, Gazetteer & Directory of Lincolnshire*, 1882, Winfrey's father is listed as R.F. Winfrey, farmer, grazier, landowner and registrar of marriages. Winfrey's first job was as an apprentice to a chemist in Long Sutton, where the family lived, then in Grantham, followed by pharmaceutical studies in London. After qualifying he worked in London for Bell & Co. Chemists on Oxford Street. Politics became a great influence in his early life, not Toryism like his father, but Liberalism. It was this background that would make him one of the leaders of his day in furthering the lot of the working labourer and the birth of the small-holding movement. He was born into a middle-class society but also saw the plight of the labouring classes at that time. His business and political career is a story of its own merit, resulting in the creation of one of the most successful newspaper and publishing companies in the UK, EMAP.

In 1887 Richard had bought the *Spalding Guardian* newspaper, started the *North Cambs Echo* in 1893, then purchased the *Lynn News*. The seeds of Liberal politics were sown in his soul while in London, later to be transplanted in the fenland soils. This and newspapers would be the platform from which he would create his future wealth and fulfil his ethical dreams for some of the labouring class. His dream was a revival of life in the countryside, a more balanced agriculture to a breed of people who had soil in their souls. He trod the political ladder to become the Liberal MP for South West Norfolk between 1906 and 1922 and later Gainsborough from 1923 to 1924. In 1916 he was appointed secretary to the President of the Board of Agriculture in Lloyd George's wartime government.

However, it was his work for the men of the soil that I wish him to be remembered for in this book, and his vision for the allotment- and small-holders of the Fens. Little did he realise that from his background in the Fens he would move into such higher circles. Friends were his forte, especially men of substance such as Lord Carrington, Sir Halley Stewart, Sir Edward Grey, David Lloyd George, Sir Henry Campbell-Bannerman and many more. As well as sitting in the House of Commons he was the unpaid secretary to Lord Carrington who sat in the House of Lords, giving him a foot in both houses when needed.

The area he was brought up in around the Wash has witnessed large areas of land being enclosed with the loss of rights for the labouring class. His heart moved out to help the servicemen who had sacrificed so much during the First World War, and improve the

Sir Richard Winfrey MP standing on the steps of his house in Castor, near Peterborough, in the 1930s.

lot of the farm workers. His vision was to change their chances in life and rejuvenate the countryside after the turbulent years from 1870 to the Second World War, when agriculture had fallen into an abyss. Indeed, land for growing food being made available for all sectors of society would become his vocation.

Fate turned its hand when he met the prospective Liberal candidate for Spalding, Halley Stewart. The two of them set off on what looked to most people at that time an impossible mission. Halley Stewart made two fortunes in his business career, in milling flour and with the London Brick Co. His greatest achievement was near his brickyards in Bedfordshire where he built the village of Stewartby, a model in social housing and conditions at that time.

Richard had worked tirelessly, on foot and in print with his newspaper, the *Spalding Guardian*, to win Halley the previously safe Tory seat. Years earlier they had both campaigned for the labouring classes and found that fewer than 500 acres were devoted to allotments and no small-holdings whatsoever. Halley's maiden speech in the House of Commons was on the Conservatives' Allotment Bill. The two men's first experiment was 10 acres set aside for allotments in 1887 at Whaplode, which created 1,600 applicants. With this amount of interest shown they set about initiating allotment clubs in many surrounding parishes to get land on a voluntary basis.

The Liberals' rally for the working man caused many safe Conservative seats to be lost in rural areas both in parliament and local government and to appease this anomaly they introduced the Allotment Amendment Act in 1890. This permitted labourers to appeal to their county council if boards of guardians failed to provide land.

After the Small-holdings Act of 1892, empowering county councils to purchase farmland and create small-holdings, Lord Carrington offered Holland CC Welbys Farm of 85 acres near Spalding where Richard established the HCC first holdings, thirty in total. These were the first obtained in England under this act. By 1897 Holland CC had bought land for allotments in other parts of the Fens and declined an offer from Lord Carrington for a further 212 acres in Deeping Fen, Willow Tree Farm. With the allotment movement at the forefront of his ambitions, Richard formed a syndicate, South Holland Small-holdings Association (SHSHA), to rent Willow Tree Farm subletting it to tenants of the association.

Halley Stewart lost his seat in 1895 to the Conservative candidate Harry Pollock. By 1902, SHSHA had taken on Cowbit House Farm and Hop Hole Farm, all belonging to Lord Carrington, which brought their holding to almost 1,000 acres. His drive to extend the allotment- and small-holders' prospects continued. In 1899 he formed the Norfolk Small-holdings Association with three other MPs, buying a farm near Swaffham in Norfolk. This was backed by Earl Carrington, Lord Hevey and others. This was to be an ideal, not just simply to bring men back to the land, but to create a new society of people who could earn a decent living from the soil, provide for old age and not to have to go cap in hand to the 'Poor Law' guardians.

Two years later, Norfolk CC entered the market for the first time purchasing 91 acres of black fen at Chapel Farm, Nordelph, near Downham Market, for small-holdings, where Richard also bought 60 acres and two cottages himself also for small-holdings. His movement in Lincolnshire and Norfolk had by now grown in acreage, morale and self-esteem, and the two associations amalgamated into the Lincolnshire and Norfolk Small-holders Association. Both his business and political career were only just beginning when he became the first Liberal MP to be elected for South-West Norfolk in 1906, which he held until 1923.

Richard grew up around Wingland, near Sutton Bridge, and Holbeach where he had seen large farms let to a few individuals, not small tenants. Most of this land had been reclaimed from the sea and enclosed for the likes of the Crown and Guys Hospital. One farm of 1,000 acres had been farmed by one tenant and upon his death in 1906, Richard persuaded the Crown commissioners to let half the land to the Norfolk Small-holders

Association. Holdings were laid out, and cottages and farm buildings were built and painted white, coining the name 'White City'.

The Co-Partnership Farms Society was formed in 1906 by Richard and in 1908 he established Syndicate Farm on 150 acres at Wingland. This was an innovation in British agriculture, being the first attempt to apply co-partnership principles to ordinary farming. The innovation lasted until 1923 when friction between the profit-sharing workers and the farm manager soured the experiment, the farm reverting to individual holdings. Part of the experiment was the Marshland and Wingland Trading Association. This company, whose shareholding was held by the members of the Co-Partnership Farms Society, supplied all the requirements for farming as well as milling feedstuffs and marketing their produce. It remained a success but eventually was sold as a going concern.

In 1936 the leases between the Crown at Wingland and Spalding with Lord Carrington and the Norfolk Small-holders Association had come to their end, a sad day for Sir Richard and Sir Halley. Both estates were under the capable management of Carter Jonas who maintained the goodwill between landlord and tenants, retaining the legacy of Sir Richard and Sir Halley.

The remarkable feature of men such as these two was that they were pioneers in this new innovation, even before the 1908 Small Holdings and Allotments Act which compelled councils to buy land for small-holdings. Hares and pheasants were often mentioned in Winfrey's speeches when attacking Norfolk squires on whether game was more important than agriculture and the agricultural worker. The Tory *Lynn News* on the occasion of his knighthood in 1914 versified this leg-pull:

> We are very much delighted
> That Richard Winfrey's knighted
> For his industry and zeal in our affairs;
> But it's much to be regretted
> That, not being Baronetted,
> The distinction does not follow to his 'Hares'. . .

The *Lynn News* and *Lynn Advertiser* were amalgamated in 1922. The *Advertiser* was a popular newspaper with the royal family when at Sandringham House.

On the occasion of the ending of the leases between Lord Carrington and the winding-up of the Lincolnshire and Norfolk Small-holdings Association in 1936, after a period of thirty-two years, a dinner was held at Sir Richard's house at Castor, Peterborough. Sir Richard remarked, 'Taking it all together it has been a glorious experiment.'

For the councils' schemes to work it had to be beneficial to both landlord and the small-holders, and not just a Utopian dream of a town-dweller moving out to the Garden of Eden. Likewise it would not be a quick fix for the unemployed. For this to work it required a new breed of agriculturalist who must have 'soil in their soul'. The Fens were the perfect area for this new form of agriculture with its rich fertile soils – it would not work on the poorer soils so often found near the industrial areas of the UK. The dream was to replace the old yeoman farmer of England with a 'new yeoman farmer' who would feed the nation during future wars if necessary, and during peacetime. Many of those early tenants achieved those goals, going on to have large commercial units today. Several had sons who have followed in their fathers' footsteps and are still on CC holdings today. The passing of time has restructured this unique system of farming and horticulture. Many councils around the UK have sold off their land; those that remain have taken the step of amalgamating holdings in order to make them more viable.

PEARCE AND SON

Alan Pearce and his son Andrew are tenants on the Lincolnshire County Council's Stowgate Estate east of Deeping St James which lies on the fen edge. The farm consists of 320 acres of all-arable cropping, which Alan took possession of in 1975, moving from a small farm near Grantham.

Their farm is part of a block of land of some 1,200 acres owned by the Lincolnshire County Council. The soil around them is very typical of fen edge type, not black peat fen, but fairly high in organic matter, lying on gravel and marine clay sub-soils, ideal root land for potatoes and sugar beet. In 1986 Alan's son Andrew took possession of 70 acres belonging to the Moulton Harrox Trust in the Moulton Chapel area. The trust dates back to 1560 when John Harrox died and left land and property to the Moulton Free Grammar School.

Soil types on the two holdings differ: Stowgate is mainly light black organic with some heavier land which is typical of that area. The Moulton Chapel land consists of silt soils, ranging from light to heavy, commonplace in that particular area. Their rotation is based on two cereal crops of wheat and barley, together with oilseed rape, peas, sugar beet and potatoes. Having six different crops gives a wide rotation for the farms, ensuring good husbandry on their types of soils. The potato crop, which annually consists of 25 acres, is the most labour-intensive and highest investment crop on the farm. The varieties grown are Cara, Picasso and Maris Piper.

Where many smaller growers have fallen by the wayside and where the Pearce family have survived, is through their marketing strategy, supplying to niche markets. Their sales from a tractor and trailer on a busy road at weekends account for 30 per cent of their production, of which 80 per cent are regular, satisfied customers. Picassos are sold from September through to February followed by Caras from February to March. Interestingly, their customer base is wide having sold to people from France, Belgium and Ireland as well as people coming to the area from London, Sheffield, Derby and Nottingham as well as other cities. At Christmas time their customers visiting the area seem to come from all over the country, and are eager to return home with a bag of potatoes from Lincolnshire.

Fish and chip shops take their potatoes, as do some wholesale suppliers, and sometimes they may even finish up as frozen chips. The venture is an example of how a family farm of moderate size can still survive through good marketing. Andy and his

Andy Pearce grading potatoes from store with his father Alan and two neighbours at Crowland Common in 2009. (RS)

Alan Pearce selling potatoes in a lay-by near Crowland with eager customers waiting in 2009. (RS)

father do most of the work, helped at peak times by local friends. Alan, like many men of his era, learnt farming on his feet, as did Andy although he furthered his farming knowledge by attending a day release course at Caythorpe College, spread over two years, culminating with a National Diploma at Riseholme College.

The smaller holdings, such as theirs, have suffered the most in modern times and many have ceased to be viable units. Had it not been for the county councils' policy of adding retiring tenants' holdings to existing holdings when they become vacant, families such as this would not have survived. This area is known as the Stowe Estate and covers 1,200 acres and in the mid-1970s was farmed by thirteen tenants; now there are eight.

Questions I asked Andrew:
RS: What were the most difficult periods in farming for the family, and why?
AP: The late 1990s and early 2000s, because of low returns.
RS: What were the best and times why?
AP: The early 1990s, with set-aside coinciding with higher grain prices.
RS: How do you see the future?
AP: Hopefully higher grain prices due to higher demand and insufficient supply. Potato prices need to increase due to increased cost of production. Direct sales to the general public allow us more control of these prices and currently the public seem to be interested in buying direct from local growers and willing to support us. Sugar prices need to rise if the crop is to be worthwhile growing. To control weeds in peas we need a contact herbicide to fall back on when residuals fail on our organic soils.

Grain and potato prices fell considerably while researching this book, but have since risen.

THE WILCOX FAMILY

The decline in farming incomes in the late 1990s did create opportunities for new entrants to step on to the farming ladder if one had sufficient capital and pluck. Two such new entrants were Alex and Joanna Wilcox who are tenants on the Norfolk CC Estate and Cambridgeshire CC Estate.

Farming prices started to decline from 1998, gradually sliding until they were the lowest seen for more than three decades. The large farms were struggling to make a profit and the smaller ones were falling by the wayside. Tenants were in dire trouble on the CC holdings with some vacating them. Since most lived in tied cottages on the holdings, many could not afford a house elsewhere to move into. The rural cottages in the Fens, once worthless, have become a refuge for the urban dweller.

Alex Wilcox grew up on a 150-acre farm in Bedfordshire where the family had a 400 sow pig enterprise. His father supplemented his income with his own spray contracting business, and also worked as an agronomist. Alex was born in 1968, the youngest of seven children. There was no future for him on the farm but his farming interests took him to Moulton Agricultural College in 1987, after which he worked on two mixed farms for six years. This put practical experience under his belt, but his ambitions were higher. He went to Nottingham University, graduating in 1997 with a degree in Agricultural Agronomy. While at Nottingham University he met his future wife, Joanna, who graduated with an Honours Degree in Horticulture in 1996.

On leaving university he took a job as an agronomist with H.L. Hutchinson, a company specialising in agro-chemicals for the farming and horticultural industry at Wisbech. At the same time he found time to marry. Walking the fen soils was not enough to fulfil their ambitions – they wanted soil on their hands as well as their boots, and it had to be their soil. Alex was at this time doing some consultancy work for the Cambridgeshire County Council and saw a 60-acre silt land holding was about to be let at Friday Bridge near Wisbech. He tendered for it and was successful, moving into the house on the holding in 2000. Able to carry on working as a full-time agronomist ensured his income while the couple ran their first farm part-time with the help of contractors.

The opportunity came again for another holding, 170 acres at Hill Farm, Stowbridge, this time with the Norfolk CC on black fen soil. Alex took this holding in 2003 while maintaining his full-time job as an agronomist covering 18,000 acres. Various other small plots of land were added to his farming operations, the latest being at Denver, bringing their total to 450 acres.

Hill Farm has become their home, as well as a farm shop where they sell potatoes, onions and home-grown asparagus, and is used as a comprehensive trials and demonstration programme for H.L. Hutchinson.

One would think that with three boys, Finlay (eight), Benjamin (six) and Arthur (four), and a farm shop to run, Joanna would have no spare time on her hands. However, while Alex is away working she does find time to load lorries with wheat and rape with the tele-handler, cart corn around in harvest time and carry out the ploughing.

The local school comes annually for a tour round the farm to learn about farming and food, and when visitors like myself call in they still have time to spare. These are the sort of people Sir Richard Winfrey fought fiercely to get on the farming ladder and he

Arthur Wilcox watches his father Alex combine harvesting on their Norfolk County Council farm in 2009.

would have been proud to see them achieving his dream and theirs. It is ironic that within a stone's throw from Alex's Hill Farm is the 60 acres Sir Richard bought in 1899. He would also have greatly approved of the way these two co-operate with neighboring tenants, Mick Denham and Barry Golding, with cultivations, harvest and any other necessities. The previous retired tenant of Hill Farm, Ron, still comes to help at busy times, and they call him 'the manager'.

WHEATLEY, TENANTS OF CAMBRIDGESHIRE COUNTY COUNCIL HOLDINGS

The Wheatley family of Coates near Whittlesey have farmed in that area for four generations. John Henry Wheatley came to Whittlesey from Saxilby in mid-Lincolnshire, then moved to Coates in 1883. After he died his wife carried on the blacksmiths shop with the help of her son, George. George did not have the best of health for the heavy work of a blacksmith so decided to take up farming. In 1911 he bought a field of 13 acres for £1,000 on the outskirts of the village to make a living on.

One of his brothers rented a council holding from the Huntingdon County Council but left home to serve in the First World War where he and another brother were killed. During their time in the war, George managed his brother's holding.

George had three sons: Frank born in 1927, his brother Don, who went into the building trade and Bert, who was killed in Crete in 1941. George was never successful in obtaining a Cambridgeshire County Council holding. He did apply for one after the First World War when his deceased brother's holding became vacant, but then ex-soldiers coming home from the trenches were given priority.

In the village there was a carpenter's shop. When Frank showed an interest in woodwork, the owner was delighted, thinking that he, being one of three boys, might come into the business. Tragedy followed when George died of a heart attack in 1943 so Frank left school at sixteen to work his father's small-holding. The soil was typical of that area around Coates – rich, fertile, black peat soil. He worked it by himself and it was not long before he added some more acres, bringing his holding up to 50 acres. His cropping consisted of sugar beet and potatoes, the rest being wheat and barley grown on a 3-year rotation, with the help of four horses.

One of Frank's proudest memories is that he had the first tractor in the village in 1948 as well as serving on the Ministry of Agriculture Food and Fisheries district committee. One field in particular he remembers having ploughed for the first time with a horse, and ever since by tractor. In 2009 he ploughed the same field for the sixty-fifth time in his life, at the age of eighty-two, a lifetime's involvement with the soil. Since the day he turned the first furrow in that field he has witnessed the gradual depletion of topsoil, now down to a few inches of peat – a harrowing experience for a man of the soil.

His son, Owen, joined him on the farm and they both built

Three generations of the Wheatley family grading potatoes in 2009. Left to right are Owen, Frank and David. *(RS)*

up a sizeable pig breeding unit of 100 sows. It was a profitable enterprise in its day, but like many agricultural enterprises, costs rose and prices declined.

Frank, like his father George, wanted a CC holding but also was not successful in getting one. The third in line in the family to try for one was Owen, who was successful in getting a 30-acre holding near Leverington with Cambridgeshire County Council in 1975. He was also farming his father's land at Coates. The next in line was Owen's son David who took a starter holding of 80 acres in Benwick, converting to a full holding of 250 acres in 2000 with a house. More land has been added, bringing their total acreage to 310 acres. With extra acres to manage, and pigs becoming unprofitable, the family closed the unit down.

The Wisbech area has always been the heart of the apple-growing industry in the Fens. Far from its former glory, it is still an important crop for many growers. The Cambridgeshire County Council holdings around David's farm consisted of many fruit orchards although there were large areas grubbed up in the 1970s when grants were available to convert them to arable cropping.

The CCC had some holdings where old established orchards had been left uncropped and neglected for many years and, with the demise of the apple industry, many old varieties were in danger of becoming extinct. The CCC together with David decided to preserve some of these neglected orchards to remedy the situation. The area in total in the Countryside Stewardship Scheme (CSS) is 28ha (69 acres) of which 9.2ha (22.7 acres) are orchards, the rest being native woodland and grass. Sheep are used to graze the orchards and grassland and David receives a payment from RDPE (Rural Development Programme for England, a conservation grant) for managing them. His initial work consisted of clearing brambles from around the old trees, followed by a programme of pruning. Pruning had to be done over several years in a gradual process to return them to their original state. Some of the orchards are thought to be 80 to 100 years old, some with local varieties, such as Emneth Early and Allington Pippin alongside the more well know varieties such as Bramley. A replacement programme has been carried out since 2004 with 80 new trees of local origin being planted.

The fruit is not harvested as a commercial enterprise but some is harvested and used for juice, marketed by a local company. Open days for the public are held throughout the growing season and people do have access to walk through the orchards. There is also an area where experiments are carried out grafting old varieties of fruit trees on to

David Wheatley in the orchards he has restored under the CCS Environment Scheme at Wisbech St Mary in 2009.

root stocks with a view to reviving them for the future. The project is a fine example of the CCC preserving a part of our agricultural heritage in the Fens as well as providing an income for David.

PHILIP MARTIN, NEWLINGS FARM, STOWBRIDGE, KING'S LYNN

William Bower, after returning from the First World War, applied for a Norfolk CC holding and was successful. He remained farming between the wars and then during the Second World War. At the latter end of the war he and many other farmers employed German POWs from the nearby Shouldham Gap camp. Like many of his generation, he had done his bit for the country once, and being in a reserved occupation made him exempt from being called up in the Second World War.

In 1924 Gotthilf Martin was born in Friedenstal in Bessarabia, a country which had experienced many changes in its rulers for hundreds of years. After the First World War it unified with Romania. There were a large percentage of Germans living in this area and with the Hitler-Stalin Pact in 1939 many were repatriated to German soil. After the pact broke up in 1941, the area came under German occupation. Gotthilf, being of German descent, moved with his family to occupied Poland where he worked on farms in the equivalent of our Land Army. He also worked in Russia for a time, returning to join the German Army in January 1943, aged nineteen. The regiment he joined was the 9th Panzer Division, which was sent to Russia in March 1944 to assist in liberating German soldiers who were surrounded by the Russian army. His division was called to France later in 1944 for the Normandy invasion and fought at the battle of Arnhem. Later that year they were re-equipped and were involved in the Ardennes Offensive in 1945. It was during this offensive that a bridge was blown up behind their lines cutting off their retreat and as a result they were captured by the Americans.

The reader may be thinking that this paragraph has nothing to do with county council holdings in Norfolk. However, Gotthilf was taken prisoner and sent to England in 1945, spending time at Moreton-in-Marsh then working at Wyndham College, Norfolk, an American military hospital. He was then transferred to Shouldham Gap camp near Downham Market from where he worked on the neighbouring fen farms. One such farm belonged to the Norfolk CC estate at Stowbridge on the holding of William Bower.

Not wishing to return to his homeland after the war, which was by then under Communist rule, he applied to stay in England, and was successful. Mr Bower took a liking to Gotthilf and when he obtained permission to stay in England, he allowed him to stay on the farm in some very simple accommodation. A relationship developed between his daughter Hilda and Gotthilf, eventually resulting in marriage and two sons, Philip and Paul, and a daughter, Anna. When he was required, he worked for his father-in-law and spent the rest of the year on several farms in the locality. He applied several times for a starter holding with the Norfolk CC but was unsuccessful until 1954, when he was given a 10-acre site. Being a German POW was not in his favour since British ex-servicemen were given priority for starter holdings.

His first crops were potatoes, sugar beet and cereals, along with pigs and poultry to supplement his income. His wife Hilda worked on the farm alongside him and in 1958 they were given another 35 acres, adding more in the 1960s, eventually farming a total of 72 acres. He never worked horses on the holding, buying his first tractor in 1954, a Fordson Major TVO, although he did eventually replace the engine with a Perkins P6 diesel engine.

A strong community spirit existed on these CC holdings, where people helped each other with work on the farms and made social relationships. Often tenants on the smaller starter holdings would supplement their incomes by working part-time on larger holdings. There was much contract work done by those who had bought larger

Above: Massey Harris 726 bagger combine harvester in the 1960s. Gotthilf Martin standing on the right, on his Norfolk CC farm on the Stow Estate near Downham Market.

Left: Gotthilf Martin filling a potato rocker on his Norfolk CC farm on the Stow Estate near Downham Market in the 1960s.

Below: Macrobert potato rocker with, left to right, Hilda Martin, daughter Betty, son Philip and dog Whisky.

machines than they themselves required. The attitude was very much to share and get on with each other, a working community at its best.

Gotthilf's tenancy was a 'retirement tenancy' so in 1989, when he was sixty-five, he retired from farming. If he had gone back to Romania after the war when it was under Communist rule, his future would have been uncertain, and more than likely he would have been shot for being in the German army. He also knew nothing of his family's whereabouts.

Most of his farming years had been fruitful albeit for the latter 1980s when costs had risen and sales declined. After retirement he stayed on in the house for a short time, before moving to his own bungalow where he lived with Hilda until she died in 2001. In 2008 he moved into a retirement home in Downham Market where I visited him to research this book. The fen soils gave him a good living and provided for him and his wife in their retirement. A remark he made to me as I left him was, 'England has been good to me.'

His eldest son Philip followed in his father's trade. In 1975 he applied for a holding of 65 acres which included a house, which was next to his father's holding. He now has 120 acres, growing winter wheat, spring barley and sugar beet and like his father works with neighbouring tenants. When asked what has been the most difficult period in his farming career he said, 'the present time,' with low prices for his crops leaving minimum margins to live on. Other regrets were that if he had known how hard it would be he would have learnt a trade like his brother, and would not have gone into farming. The soil has changed even in his time, with loss of topsoil and, as a consequence, poor drainage in the fields which need redraining with field drains. He ploughs about 10in deep which is almost in to the clay sub-soil. Another concern he has is where he will live when he retires at the end of his retirement tenancy. Maybe, unlike his father, he did not have 'soil in his soul'?

Although the council holdings nationally have declined, the fenland estates have not declined at the same rate. This could be due to their soil types being above average, and maybe the tenacity of the fenmen themselves, a hardy breed, the last of the 'new yeomen'. The loss of acreage may have stabilised, but there are now fewer new entrants, not through lack of incentive, but through the amalgamation of holdings. The past two decades in the Fens have witnessed the arrival of larger agribusinesses which own and hire land for farming and growing crops. The middle-sized farmers as such have declined, who now through contract farming agreements and farm business tenancies are letting their land to the larger growers.

The smaller farmer/grower, however, has survived in the Fens, catering for specialised and niche markets, many of which are using CC holdings. We are the last enclave of small farming in the UK with more than 40,000 acres still under the CC system. Inevitably, the size of holdings will increase, economies of scale will dictate that facet of farming. It is ironic that the policy of breaking up large estates 100 years ago is turning full-circle; history always repeats itself, especially in farming. The future of this system of land management will not be determined by the industry itself, but by the politics of governments and councils, as it was when it was conceived. Since the Small-holders and Allotments Act of 1908, farming has witnessed many changes, good and bad, especially in the Fens.

I believe that for the tenants to survive and prosper on all CC estates, there will have to be more joint ventures between the tenants themselves. If sufficient returns for the CC estates from these land holdings is not maintained, outside forces will intervene. The CC should not, however, forget the foresight of such fenmen as Sir Richard and his band of idealists. How many county councillors even know of the 1908 act? Should not the centenary of this important act have been recognised and celebrated to honour the pioneers who conceived it?

These estates have helped feed this country during two world wars, conceived many farming successes, and the land is still there unlike many investments these assets might have been exchanged for. It has been a dream fulfilled, a reality in human relations with the soil and the last area of agriculture where there is an opportunity to become a farmer and grower. In 1908 it became law for county councils to provide allotments and small-holdings for their residents, it is now above the law that they should maintain this national heritage, and not become a chapter in our history books.

County Council Estates:

Cambridgeshire:	34,007 acres	Estimated fen: 14,000 acres
Lincolnshire:	17,578 acres owned, 2,485 leased	Estimated fen: 15,000 acres
Norfolk:	16,604 acres	Estimated fen: 9,000 acres

The total area of County Council farmed land is approximately 70,000 acres of which I estimate 38,000 could be classed as fen and fen edge.

THE BEDFORD CONNECTION

The Dukes of Bedford had, since the creation of the Thorney Estate, always provided with extreme consideration for all sectors of society on this estate. Long before the Allotments Extension Act of 1882, the labourers on the Thorney Estate had gardens to meet their own requirements. After the act, the existing gardens were deemed too small and were enlarged. The duke was somewhat restrictive on some of the allotments, specifying that holders could only cultivate out of their working hours.

Before the Small-holdings Act, empowering county councils to purchase farmland and create small-holdings, was passed in 1892, the Duke of Bedford's estate in Thorney had already created four small-holdings on Lady Day (25 March) 1889. In 1894 four cottages were added to these holdings. In the same year a government act was introduced advising councils to take over the management of allotments from existing landlords. The Thorney Parish Council declined this offer.

> The Thorney Parish Council is of the opinion that, could the desire for larger holdings be met by the letting of land for such a purpose in some suitable and convenient place, it would be advisable for all allotment holders to become direct tenants of His Grace, rather than the Parish council.

Thirteen councils in the Beds and Bucks counties on the Woburn Estate also declined the offer. An indication of the dire state of agriculture at this period is the remissions in rent for these allotment holders. The rent for the cottages remained the same but the rent for the land was reduced by: 1890 – 25 per cent, 1891 – 12.5 per cent, 1892 – 32.5 per cent, 1893 – 37.5 per cent, 1894 – 50 per cent and 1895 – 25 per cent. His Grace did state that for the tenants, these holdings have been a success, had the land remained in one holding the rent would have only been 1s per acre less. The estate however did have to consider the burden of added investment in fences and buildings to make them viable for the tenants.

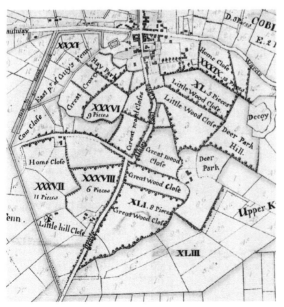

Map of 1756 showing Thorney village which was the hub of the 20,000-acre estate. (By kind permission of the Marquess of Tavistock and the Trustees of the Bedford Estate)

Draining of the Great Levels in the seventeenth century was the Bedford family's passion. Developing their Thorney Agricultural Estate during the eighteenth and nineteenth centuries was their vocation, and leaving us with a village and farmsteads was their cenotaph to this great family.

In the 11th Duke of Bedford's book *The Story of a Great Agricultural Estate*, published in 1897, he says, 'Only 300 acres of culturable land came to the House of Bedford on the dissolution of the monasteries.' A note on the same page from Warner's *History of Thorney Abbey* says a document is quoted, endorsed in the hand-writing of Elizabeth I's illustrious minister Cecil, stating 17,760 acres belonging to Thorney Abbey. This figure is somewhat close to the acreage in 1909 when the estate was sold.

After the initial draining in the seventeenth century, the land around Thorney began to be developed as an agricultural estate by the 5th Earl, 1st Duke of Bedford (1613–1700), and this was continued by successive family members.

The 5th Duke was the first of the line to make his mark in the agricultural field, taking a keen interest in all aspects of the developing industry at that time and even becoming President of the Smithfield Club. The 6th Duke followed in his footsteps and became the first Governor of the Agricultural Society in 1838. He, along with the 7th, 8th and 9th Dukes, developed the land around the village of Thorney into a model fenland agricultural estate during the nineteenth century. This was during a period when huge drainage schemes were being carried out, both in the Fens and the river estuaries.

The invention of steam beam engines, driving first scoop wheels then centrifugal pumps, would finally turn Thorney Estate into some of the best farming land in the Fens. New sluices were erected and many drains cut, employing the finest engineers such as Telford and Rennie. The duke's agent, Tyco Wing, who was influential in the Nene outfall of 1830, stated in evidence before a committee of the House of Lords in 1848 that 'the value of land in some parts of the North Level had increased 100 per cent, and in some cases more, and it is well known that land that might have been bought twenty years ago for £5 per acre, would now bring from £60 to £70.' The tenants on the estate had experienced a somewhat stable income, which encouraged the Bedford family to invest in the estate's infrastructure.

This was the time when wealthy landlords were building new farmsteads as well as houses and cottages of exquisite designs throughout the land. It could be said in many ways it was the renaissance era of agricultural buildings. The Thorney Estate stood out from all other estates in the Fens where landlords had carried out similar works. No other private landlord had such a large estate in the Fens as at Thorney. The Duke of Portland's estate at Postland, less than half the size of Thorney, was developed during the late nineteenth century but did not compare in grandeur. Portland's estate was purchased from the Marquis of Exeter who, while holding the lordship of the manor of Crowland, did not develop it in the way the Bedford family developed Thorney.

What did make Thorney stand out from other fenland estates was its village, being in the centre of the estate and owned totally by the Bedfords – this is unique in fenland history. The agricultural depression, commencing in the 1870s through to the early twentieth century, coupled with poor harvests took its toll on the estate. It was a time also of a new era in farming with the creation of allotments and small-holdings of which this family were at the forefront.

The acreage of the Thorney Estate from 1816 to 1895 was between 18,000 and 19,000 acres. Rents from the farming estate between 1816 and 1821 remained constant at £21,000, then fell to £17,000 in 1827. From that date they rose steadily to 1878, peaking at £38,000. This was in the first period of the depression years and from then to 1895 the rents fluctuated between £30,000 and £18,000.

The 11th Duke of Bedford in his 1897 book says, 'I am told on excellent authority that the Thorney Estate, and part of the Bucks and Bedford estates, are at present practically unsaleable.'

Herbrand, 11th Duke of Bedford
who sold the Thorney Estate
in 1909. (Reproduced by kind
permission of the Marquess of
Tavistock and the Trustees of the
Bedford Estate)

The 11th Duke, encumbered with death duties and declining rents from his Thorney Estate, was attracting interest from several quarters. In 1902 he refused an offer of £750,000 for the estate and in 1907 refused to negotiate for the sale on an offer of £800,000. Both offers were believed to have been made on behalf of land speculators. Interest was also shown by the Crown to buy a part of the estate via Lord Carrington who proposed that the duke should raise the rents to a commercial level before the sale. The duke consented but told Lord Carrington he would reimburse the increase in rent to his tenants after the sale – the negotiations fell through.

The estate was finally sold in 1909. It was one of the most remarkable land sales for its size ever transacted. No sale particulars were drawn up and the whole transaction was carried out in privacy and goodwill. The entire village was also sold, wherever possible, to existing tenants. A local newspaper reported on 26 June 1909, 'It was never doubted the duke would treat his tenants with the consideration he has always shown towards them, and they were greatly relieved when they knew that the occupiers of holdings outside what was known as the "Crown area" had been given the chance to purchase their holdings.' Many tenants bought their freehold, others did 'back to back deals', buying their farms then selling to another landlord while retaining the tenancy.

A quotation from the duke's book reads:

The only pleasure which I and my forebears can have derived from Thorney is the kindly feeling which has existed between us and our tenants, and the inhabitants of Thorney town. As an instance, I mention that there has never been an eviction of a farm tenant.

5

Reclaimers

The Wash sea defences were first constructed by man possibly as early as the Roman period. Many historians refute this theory and indeed say they were built much later, possibly during the great monastic era. The banks which do exist are drawn on our OS maps as Roman banks and may have been on the lines of Roman banks, giving them their name. Whoever built them, whether Roman or Anglo-Saxon, we do know that these banks were built more than 1,000 years ago.

Nature itself has also assisted in providing sea defences by a gradual build-up of estuarine mud, thereby extending the 'green marsh area'. This is an area between the dry land and the area of mud that is constantly under salt water. Because this area is only under salt water during high tides, salt-tolerant herbaceous plants establish themselves, feeding on nutrients washed in by the tides. Left to nature, these areas gradually extend themselves seaward, which has happened around the Wash over history.

The alluvial matter deposited in the Wash is almost entirely sourced from the rivers flowing into the Wash itself, gathered from their catchment areas. Alluvial deposits have been the greatest on the southern section of the Wash where the Witham, Welland, Nene and Great Ouse flow into the delta, though they have also been found to a lesser extent on the west section and even less on the east section. As this occurred, mankind built banks on the edges of the green marsh excluding the intrusion of salt water and in time this land has been put to agricultural use.

These sea defences can be seen around the Wash from Skegness, Boston, and Spalding, across to the Great Ouse in the west of the Fens. Some are further inland than others and date from the Roman period to the twentieth century. Notable dates of building and expansion were 1280, 1660, 1793, and 1810.

Courts of Sewers were established in the thirteenth century by royal consent with commissions drawn up to look into reducing flooding in various parts of the Fens. They could order landowners to maintain and restore banks to their original condition but not to improve them. Consequently, no advances were made in sea defences until the seventeenth century.

Up to this period most of the embankments that had been made were mainly for sea defences and not reclamation. Prior to this time many areas were kept outside the sea defences for salt production, which necessitated the influx of salt water. By the sixteenth century salt production had virtually ceased on the marshes and drainage in earnest had begun in the fenland interiors. Adventurers were looking to reclaim land from the sea as well as the low-lying land away from the Wash. The eighteenth century saw many sea defences built around the Wash basin, from Spalding to King's Lynn.

After a damaging tidal surge in 1810, the Courts of Sewers were given permission by parliament to raise taxes to construct a new sea bank along the entire western section of the Wash coast between Boston and Skegness. This became known as the 1810 bank and remained the main line of sea defence well into the twentieth century. It can still be seen today, all inland from the present coastline, except a section near Wrangle.

The nineteenth century was the time of extensive sea defence construction and enclosures, with many such works carried out mechanically with steam-powered machines at the end of the century. The twentieth century witnessed reclamations on the western Wash coast between Wrangle and Skegness, Wingland, and north of King's Lynn. Another interesting reclamation was on the Witham estuary at the North Sea Camp, which is referred to later in this chapter.

Two world wars in the twentieth century highlighted the vulnerability of this island nation to sustain food supplies. Although land drainage across the Fens and marsh was a priority before and during the Second World War, no reclamation was carried out at that time but began in earnest during the post-war years. The first was a major intake in 1948 along the Lincolnshire coast north of Wrangle, with work involving almost 6 miles of new sea bank where several local families were involved, notably Roughton, Atkin, Worth and Atkinson. This linked up with the 1920 intake carried out by the Worth family at Wainfleet. A small intake near Wrangle in 1960 was carried out by H.M. Clarke, followed by three more intakes along the line of the 1948 intake in 1966, 1973 and 1976. South of this area, four intakes were done by the North Sea Camp between 1952 and 1979 and these joined up to the one carried out in 1972 by the Crown Estate and John Saul.

The Ward family enclosed a large area along the Welland in 1949, followed a year later by the Hays and Thompson families. On the Norfolk shoreline of the marsh between Sutton Bridge and King's Lynn, four intakes were created in 1951, 1955 and 1974. North of King's Lynn five intakes were carried out between 1950 and 1967. All the intakes during the twentieth century along the Norfolk coastline were carried out by the Crown Estate which also did one on the Lincolnshire coast. The rest were carried out by private landowners.

Environmental objections, if they were voiced, carried little weight during that period of reclamation, and planning permission for such projects did not exist. Tax incentives were in place to offset the capital works against income tax on work done during the early post-war years. Later, in the 1970s, government grants were available to fund projects through MAFF and the EU with FEOGA grants.

The drive for home-produced food was dwindling by the 1970s and reclaiming salt marshes for agricultural land was seen as hardly cost-effective. These factors were also seen as detrimental to the marsh environment. Conservation, not reclamation, was to become a major issue which resulted in Norfolk County Council imposing a planning moratorium on future reclamation, which was followed by Lincolnshire County Council in the 1980s.

Since Roman times it is estimated that between 70,000 and 75,000 acres have been reclaimed from the Wash area: 35,000 acres during the seventeenth century, 19,000 acres in the eighteenth century, 10,000 acres in the nineteenth century, and 10,000 acres in the twentieth century. This equates to approximately 10 per cent of the total area of the Fens, but more importantly it is almost all Grade 1 agricultural land designated as some of the best soils in the UK. Wheeler, in 1896, estimated 63,299 acres had been reclaimed.

In the years after the Second World War, concern was mounting with regard to the ever-growing demand for fresh water. In 1972 consulting engineers Binnie & Partners were appointed by the Water Resources Board to carry out a feasibility study of constructing fresh water reservoirs on the fringes of the Wash in the tidal section. They would have been filled with fresh water from the Rivers Welland, Nene and the Great Ouse. The area between the existing sea defences and the reservoirs would have eventually silted up and reverted to green marsh with the potential of reclamation and converting to farmland. The scheme did not materialise.

LOGISTICS OF RECLAMATION

There are many factors to take into consideration when proposing to reclaim saltings from the sea such as its economic viability (compared to the price of farmland in the area, if available to purchase), the acreage of saltings and the quality of the land to be reclaimed. The land can be too heavy or too sandy due to action of the tides on the foreshore, or the length of time the saltings have been in existence. Bank construction is of the utmost importance. The area to be taken in must warrant the cost of the bank construction and large oblong or square areas are most favourable. The soil available to build the bank must be of the correct consistency – too much sand results in a bank with no body, easily eroded by heavy seas, and too much clay results in slipping material with no shear strength when saturated. Account must also be taken of the amount of protection that is likely to be offered by the remaining saltings against wave action, tidal surge and the prevailing wind direction. It is not advisable in the Wash to construct the bank beyond the edge of the saltings. The time of year when bank construction is carried out is also of vital importance. This work should be carried after high spring tides, and before high tides in the autumn, so between April and August is the optimum time. Most of the problems with bank erosion have been along the western sea banks between Boston and Wainfleet.

NORTH SEA CAMP

To include one of Her Majesty's open prisons in a book on fenland agriculture may seem somewhat out of context to the reader. However, this prison has been intrinsic in the reclamation of land. In the years after the First World War, there was an increase in the birth rate, which in turn had an impact on the elementary schools' ability to manage so many pupils. As a consequence there was a sharp rise in numbers entering approved schools and borstals. The authorities foresaw this problem and conceived the idea of open prisons, where boys could be involved in constructive projects, benefiting themselves as well as the country.

The site for the first experimental camp was Lowdham in Nottinghamshire. Sir Alexander Paterson MC, the prison commissioner of a borstal at Feltham in Middlesex, addressed his staff and boys as 'his pioneers', informing them that he intended to march them to Lowdham to establish a camp near the River Trent. This camp was so successful that the Prison Commission decided in 1934 to expand this form of detention on similar lines to those at Lowdham.

As far back as 1914 the Prison Commissioners had investigated the possibilities of reclaiming salt marshes on the Earl of Leicester's Holkham Estate on the north Norfolk coast near Wells-next-the-Sea, using borstal labour. The Chief Drainage Engineer investigating that project was Captain Roseveare (Land Drainage Branch of the Ministry of Agriculture). After investigation the site proved unsuitable due to the possible interference of the waterways to Wells harbour, and the First World War curtailed any further development of similar schemes.

In 1919/20, Captain Roseveare was called in to supervise the reclamation of 338 acres on the Wainfleet out-marshes south of Skegness owned by the Worth family. This was financed and developed by the Board of Agriculture when the authority's Land Reclamation Branch policy was 'to provide work for unemployed labour and demobilised soldiers on undertakings of national importance in connection with schemes for land settlement.' The original intention when the first phase was completed was to reclaim a further 660 acres to the south near Friskney, but the scheme was abandoned.

In 1932 Mr H.S. Rogers (Surveyor of Prisons) was commissioned to carry out a report on Land Reclamation by HM Prison Commission 'to endeavour to find some scheme which would be of national value, and would provide continuous and strenuous work for a considerable number of borstal inmates or certain classes of ordinary prisoners.' Captain Roseveare's experience at Wainfleet led him to suggest to Mr Rogers that he should investigate other sites along the coast between Boston and Wainfleet, which they both did in March of that year.

The report contained an ambitious scheme not seen along this section of the Wash since the 1810 embankment. It envisaged enclosing a total of 2,450 acres between the Boston Haven and the last enclosure at Wainfleet, a distance of some 15 miles. The original proposal by Rogers favoured the reclamation of the 660 acres at Friskney, which Roseveare had not seen to fruition after his successful work bordering it. One of the main stumbling blocks was to find a suitable site for a camp to house the prison staff and inmates. It needed to be isolated, yet close to a town or large village and within access to electricity and piped water.

As a result of his report the prison officials visited some of the marshes suitable for this project and decided on the site called Clayhole House to build the camp, and this would become the hub of the adventure. The site chosen was near the village of Freiston on the Witham estuary, situated approximately 5 miles south-east of the town of Boston, on the edge of the marsh lying in one of the most inhospitable parts of the Fens.

The site had foreshore on its eastern boundary, the Witham Haven on its southern boundary and on the western boundary was an area of land that had been reclaimed in 1872. It was an ideal site to commence this experiment in prison reform and the location was synonymous with the term 'open prison'.

The landscape is bleak but not devoid of history. A Roman sea bank – which began from Boston and followed the Haven then turned north where the camp is today – went on to Wainfleet, an important Roman port in its day. Ships pass by the camp sailing up and down the River Witham to and from the port of Boston.

A chosen number of boys were prepared for their march from HMP Stafford to Freiston on practice marches around the neighbouring countryside. The experience gleaned from the Lowdham Grange open prison pointed to taking lads for the new enterprise direct from the collecting centre at Wormwood Scrubs and having no lads who had been at other borstals. This limited the numbers suitable for this experiment, usually about half a dozen each month, and so for the next project it was decided to bring staff and lads together at the prison at Stafford.

Borstal boys on the march from Stafford to North Sea Camp in May 1935. They are believed to be seen along the river Trent at Nottingham. *(HMP)*

The person on the left is believed to be Major W.W. Llewellyn when he established the first base camp at Freiston in 1936. *(HMP)*

On 23 May 1935, under their leader Major W.W. Llewellyn and his staff, they set off on their march. Some reports mention twenty boys, others eighteen. Whatever the numbers, they were pioneers embarking on a great adventure, an experience that must have remained in their minds for the rest of their lives. The journey took them to Uttoxeter, Derby, Nottingham, Bingham, Grantham, Sleaford and Boston – a distance of 110 miles – taking eight days to walk. Lodgings were provided at village halls, church halls and accommodation found by the international charitable organisation Toc H. They arrived on 31 May 1935 on a perfect spring day. On arrival, a central hut was being erected on-site containing the dining hall, kitchen and offices, while their living and sleeping accommodation was a tented camp. Work began immediately on erecting the remaining huts with the help of skilled local tradesmen.

By July 1935 the number of inmates had grown to sixty-six and for the rest of that year all their efforts were utilised in providing full amenities for the camp ready for the serious work of reclamation. The intention of the operation was to create a farm from the out-marshes by building sea defences to hold back the tides. This commenced in March 1936 and finished in 1979.

Machines were available at the time to carry out those works, but were only used in the last intake in the 1970s. On the initial intakes soil and mud were dug by hand, transported in the early days by man-powered tucks on rails to form the banks; later small engines were used to tow the trucks. This creation will be remembered as man's last great conquest against the elements to create land for farming in the Fens.

Moving soil by hand labour in the late 1930s. *(HMP)*

Soil to construct the sea banks was moved on narrow gauge railway tracks pushed by hand. The green marsh can be seen in the background in the late 1930s. *(HMP)*

As mentioned, the last intake was completed in 1979, using machines to construct the bank. It was the last project to take place around the Wash. The site was not 'ripe' for reclamation when it commenced, and maybe should not have been done. There were insufficient saltings remaining after the bank was built and erosion did occur to the extent that the bank may have breached in time. A decision was made by HMP to sell this enclosure to the RSPB for a bird reserve, who pierced it in 2002 allowing sea water to flood the land.

The 1965 and 1970 intakes are in permanent pasture, grazed by the prison's flock of sheep – all being part of the reserve owned by the RSPB. The prison farm at its peak of some 600 acres was a highly productive farm producing the complete range of crops as were grown on neighbouring farms along with livestock enterprises. Much of the produce went to supply other prisons with produce.

Managed retreat and conservation are wonderful terms when times are good and food hampers full. Forced labour was indeed used, the alternative being a cell in a closed prison, but many did get great satisfaction from their work and left a worthwhile legacy for whoever was to manage it in the future. The 1970s became a decade of food surpluses in Europe, and the awareness of the declining areas of salt marshes around the UK. Conservation took over from reclamation, and the gradual demise of North Sea prison farms. This must be seen as a milestone in Fenland history when man had eventually yielded to nature for the first time.

WARD, HAY AND THOMPSON

All intakes in the twentieth century on the north Norfolk coast between the Sandringham Estate and the River Nene were carried out by the Crown Estate. The intakes on the southern section of the Wash along the Spalding and Holbeach marshes were carried out privately by three families after the Second World War.

Jo Ward, Alex Hay and George Thompson farmed a stretch of land along the foreshore in Holbeach Marsh where no previous reclamation had been carried out since 1793. That enclosure brought in many thousands of acres of farmland along the entire Wash shoreline, from Spalding to Sutton Bridge. The old sea bank, which dated back to 1660,

Left to right: J. Ward, Mr Taviner (engineer), Stuart Hay (boy) Alex Hay, George Thompson, standing on the new sea bank at Holbeach Marsh in 1948.

remained as a secondary sea defence, much of which is visible today. The three families started looking at the project in 1947, with Mr Taviner, a drainage engineer to some of the local IDBs (Internal Drainage Boards). Alex Hay had a friend in Holland who had experience of reclamation work there, so organised some of them to go over to research the work being done.

Jo Ward was the first to enclose his section in 1948, which was along the Welland estuary bordering the Hay/ Thompson proposed intake. Even though the Ward intake had a bank on their western border, another bank had to be built to allow water discharge into the Wash from the Holbeach River. The same applied to their eastern boundary where drainage access had to be left open between their intake and the 1938 intake. Their combined operation involved 670 acres, split between Hay's 490 acres and Thompson's 180 acres. It was funded by both families, pro rata, with no government grant aid, but with tax incentive schemes. The programme started in April 1949 with the deadline completion date prior to the high tides in October.

Work progressed well until the first high tide came in September when there was still 1,100 yards of bank to complete, allowing the tide to enter through the gap. Work carried on with the new bank built to a height of 19ft ODN (ODN is the Ordnance Datum Newlyn of mean sea level).

In February 1951, during a high tide, the bank was breached and water flooded into the new intake. This tide was not scheduled to be a particularly high tide but a surge down the North Sea made it so, causing the damage. Machines worked quickly to repair the breach before the next tide was due by building a new bank inside the intake to stem

This dragline was marooned when the first high tide came with the bank not completed in 1949.

Filling and laying sand bags along the tidal section of the new sea bank to stop erosion at Holbeach Marsh, after being breached in 1951.

An unexpected tidal surge down the North Sea breached the new sea bank which was repaired by three draglines before the next incoming tide in February 1951.

the tide. The next year the entire bank was increased in height to 22ft ODN, the works being funded by the government, as now it was deemed a primary sea defence. The entire sea defences along that coast were raised again to 25ft ODN.

The normal practice after reclamation on these new intakes was to graze them with cattle, which they did for about seven or eight years. They were advised not to attempt to move soil to fill creeks in before this time or until most of the sodium clay had leached out of the soil. Infilling creeks had to be carried out with caution so as not to bury too much of the topsoil, which contained organic matter. Constant monitoring of mineral elements and sodium was carried out on a regular basis, while gypsum was also applied.

Some of the soils turned out to be heavier than expected, while some were very much lighter, almost sand. The heavy soils were later covered with 4–5in of fine silt which was mined very close to the intake, improving their working and cropping potential. The lighter soils were closer to the Holbeach River discharge areas where fine silts had been deposited, while the heavier soils were away from that area.

The final cost of the new lands was around the market value at that time of £200 per acre, supplemented by 10 per cent building tax allowance, written off over ten years. Both families had no regret as it allowed them to extend their estates on their own doorstep when it would have been impossible to have done it two decades later.

Much of the soils on farms in that area had developed potato eelworm, and with no chemicals at that time to solve the problem, virgin potato land would have been an added bonus. No reclamation has been done on this section of coastline since.

Alex Hay died in 1958 leaving two sons and a daughter: George, Stuart and Mary. Stuart and George were only young boys when the reclamation was in progress and it was a wonderful experience for them to see the farm extend into the marsh. It would be Stuart who would eventually farm the new intake after he and George demerged their farming enterprise. It became a passion for Stuart to turn it into a productive part of his farming operations by improving the soil with drainage, roads and conservation habitats. The destiny of our lives can be as uncertain as the tides in the Wash, and like our crops there is a seeding and a harvest. Beginnings and ends, life and death are an everyday part of farming life. Uncertainty is a word farmers are very familiar with and no one knows this more than Mary Hay, Stuart's wife. Mary became a part of the Hay dynasty in 1966. They met when Stuart had gone to Holland to glean from the Dutch their expertise on onion growing. Mary's father traded in daffodils and she helped on the administration side of the business. In those days the trade in daffodils and other bulbs was mainly one-way traffic, from Holland to the UK. Stuart was at that time farming with his brother, George, in Holbeach and Spalding marsh, taking over from their father after he died.

A turning point in Stuart and Mary's life was in 1983 when the brothers demerged their farming enterprise and Mary took an even keener interest in the farm. The busy periods on a farm such as theirs with so much diverse cropping during the year have their exciting as well as testing times. Time schedules do not always blend into harvest operations and too often meal times revolve around what is happening on the farm that day. There is probably more agronomy, prices, weather and other farming news spoken and listened to around the farmer's kitchen table than in his office. For that reason alone, most farmers' wives know 'what is going on', even if it is only gleaned from a distance from the kitchen table, or the Aga. Mary was brought up on a farm, but not as isolated as theirs in the marsh. It was in many ways not unlike her native land, flat and reclaimed from the sea with many thousands of acres of tulips and daffodils in the surrounding area. With three daughters, Karen, Fiona and Julia, a large acreage of some

A caterpillar drags a plough through reclaimed land in the late 1950s.

Mary Hay standing next to a pit where silt was mined for improving heavy soils, now a conservation area, 2010. *(RS)*

of the best land in the marsh, life was extremely good to them. Stuart had taken on a new farm manager, Nigel Patrick, who not only fitted in well with the family but had the ability and knowledge of managing a large complex farming organisation.

Unfortunately, life is not always as planned. Stuart was diagnosed with cancer, and the fortune of life ended for him on 2 March 2000, when he died age sixty-three. As one would expect at his funeral we sang 'We plough the fields and scatter'. Farming is a series of beginnings, changing daily, wheat tillers, flowers bloom and potatoes fill the rows, mother nature is never still, everything on a farm is ongoing whatever happens in the cycle of our lives. Mary knew that Stuart's soul was deeply rooted in their farm. I asked her when she knew she would carry on the farm after Stuart died? The reply was instant, 'Both myself and the girls had no doubts about carrying on where Stuart had left off. What else would I do? We are farmers.' The farm has not stood still since that momentous decision by the family. The acreage farmed has grown since then from 2,350 acres to 2,830 acres today. Their cropping is varied to suit the farm consisting of; 1,000 acres of wheat, 120 acres of daffodils, 250 acres of vining peas, 540 acres of sugar beet, 170 acres of onions, 270 acres of potatoes and 380 acres is made up of grazing, fallow and land in the Entry Level Scheme (land under the government conservation scheme).

It was through the reclamation that Michael and David Thompson introduced cattle and sheep to their farming enterprise, as did the Hay family. The Hays went out of sheep in 1970s but the Thompsons carried on their livestock even when the reclaimed land became arable. This practice has remained with them through to David's son Richard, and they stand out as one of the very few farming families who maintain a mixed farming enterprise in the marsh.

Michael Thompson, with whom I researched this subject, passed away before I finished the book. I attended his funeral and recall the story told of his love of the marsh, and how he loved to take his nephews and nieces down to the creeks on a high tide and swim with the seals. I dedicate this section of my book to him, a gentleman, a farmer, a conservationist, a countryman and an adventurer, and above all a Marsh man.

The Crown Estate/John Saul Reclamation on the Western Wash Shoreline

This intake involved 500 acres, being smaller than the Hay/Thompson project, but would require almost 1 mile more sea bank. Speed of construction was paramount to its success as the bank was built between the neap tides of April and September 1971. The topsoil was kept separate to put on top of the finished bank as the excavated soil was too 'raw' and unstable, being unsuitable for rapid grass establishment which is vital to protect the bank from wave action. Three draglines were used in a team, the smallest furthest out to sea passing its bucket full to the second, medium-sized one which passed to the largest dragline which put the soil from the three draglines onto the area of the new bank. After depositing the soil, bulldozers pushed it into shape to form the new bank. All soil for the bank came from the sea side of bank leaving a burrow pit, which silted up to ground level within ten years. This is how the present sea defences could be heightened today with all soil needed being on site, should the need, or intention, arise against rising sea levels.

After enclosure, dykes were dug to create drainage for the fields and the soil from that operation was used to fill in the creeks. No soil whatsoever was brought into the intake area from outside, eliminating the risk of contamination of soil-borne diseases such as potato eelworm and injurious weeds.

In preparation for future arable cropping, the land was levelled using a crawler tractor and disc harrows for sheep to graze. It was estimated that 100in of rain (4/5 years' rainfall) was required to reduce sodium content enough for cropping. After that period the land was under-drained with pipes for field drainage. This intake did turn out to be of heavier soils than the 1810 intake, although no problems occurred on the banks

Left to right: Tobias, Dulcie (John's widow), Oliver, Michele and Archie Saul standing in front of the monument to John Saul in May 2010 who built the bank in 1972.

Combine harvester working on reclaimed land of the Saul family in 2008, out marsh is in the background behind the sea wall.

with wave erosion. The 1976 intake which was north of the Saul intake, has suffered bank erosion since its construction, especially in the tidal surge of 1978. It is suggested that insufficient green marsh was left in front of the bank when built.

The Environment Agency are preparing a report on sea defences along that coast and it is mooted to be moving towards a policy of breaching some outer sea banks and improving inner banks – hoping that the area between will accrete before sea levels rise. It has been customary for the sea banks built during the past century to be adopted by the Environment Agency and the land enclosed to be also be adopted by the IDB of that area. There are, however, sections of bank still not adopted where problems have occurred due to erosion of the bank.

WILL WE EVER SEE MORE RECLAMATION OF THE WASH IN ANY SHAPE OR FORM?

Peter Dawe has formed the Wash Barrier Corporation Plc, which has drawn up proposals to construct a barrier across the mouth of the Wash. The area of green marsh in the Wash is growing, especially along the southern area through siltation, and where previously fresh water from drainage dykes ran into the saltings by gravity flow, the water now has to be lifted by pumps into the same area. If a rise in sea levels does not occur one would assume the Wash in time could become almost entirely green marsh except for the channels taking the water from the Fenland rivers.

The term 'Adventurers' is more than a folklore in the Fens – they were speculators who drained the four levels in the fen interior in the seventeenth and eighteenth century. In many ways, reclamation from the sea in the following centuries, by the Crown Estate, marsh farmers and the pioneers of North Sea Camp was a reincarnation of that era. For the young inmates it was not their choice, it was their fate. All involved have left a lasting legacy as long as man requires that land for his own use, for farming or conservation. A part of every human soul who worked on this enterprise is embedded in those sea walls and dykes. For the farming families it was a challenge and a legacy for future generations to farm soils taken from the sea by their forebears.

6

Those Who Toil – Agricultural Labour

Manual labour has always provided the engine of the great agricultural powerhouse that is the Fens. Its workforce has been ever-changing from when the Romans first set foot here to the influx of Eastern European migrants working here today. Many people of different nationalities, different religions, male and female, and of all age groups have trodden these fenland soils. All have toiled to drain, to plant and to harvest its crops and to prepare and process them for the general public. Its soils are impregnated with a history of people, labouring people, whose genes mingle with its fine particles of earth not unlike the layered roddons across this land.

Over the past 500 years it has been the fen soils that have drawn people here. Huguenots, Walloons and Flemings fled from Europe because of Catholic persecution during the seventeenth century. They were drawn here by the Russell family who were the Earls of Bedford, their task: to create a Fenland paradise and then farm it. With more land in the Fens drained, a larger populace of labour was required during harvest periods. Cole seed for oil was introduced by the Dutch settlers, while woad was grown for using in dyes, especially in Tudor times. Until the mid-nineteenth century these crops, as well as cereals and mustard, were mown by hand, tied in sheaths and put into stooks in the fields to dry. After drying in the stook they were taken by horse and cart or wagon to the 'stack yard' or barn to await thrashing. In addition, rich pastures on the fen and marshes produced fine cattle for the butchers, and sheep were one of the main livestock enterprises providing fine wool for the Flemish weavers. The Huguenots who came to

A gang of men harvesting in Thorney Fen.

the Fens were mainly from middle-class families drawn from farming stock trades and professions. Few if any were labourers.

The end of the Napoleonic Wars in 1815 marked the cessation of hostilities with France and the end of the demands that the war made on our economy. This also coincided with the time when much of the fen and marsh was being drained and enclosed for arable farming. Pastoral farming, however, was still the main agricultural enterprise in the Fens and marsh until this time. The introduction of steam engines for draining, cultivating, pumping and transportation, combined with the two world wars in the twentieth century made arable farming dominant, with the resultant requirement for a larger labour force.

Many of the fenland estates restructured their farms and farmhouses and buildings during the mid- to late nineteenth century. Drainage pumps, bridges, culverts, slackers and sluices were also updated to improve the farms for arable cropping. Hundreds of farm labourers' cottages were built of brick improving the lot for the workers. Even the remaining areas of old pasture had cottages built on them allowing the stockmen to live 'on the job'.

THE IRISH LABOUR FORCE

Extracts from *Fenland Notes & Queries*, published 1892–1909, give mention of migrant Irish labour in the Fens around Wisbech. In 1736 the parish burial register of Elm, near Wisbech, says that three labourers were buried in paupers' graves, all unnamed, in each case being referred to as 'a labourer from Ireland'. In 1744 another burial was named 'James . . . an Irishman,' who were all 'wandering harvestmen'.

On Friday 3 July 1795 four Irishmen, Michael Quinn, James Culley, Thomas Quinn and Thomas Makin murdered William Marriot at his house in Wisbech Fen. John Letts of Guyhirn went off in pursuit of these men and caught up with them in Uttoxeter, where they were arrested. They were brought back to Wisbech, tried and executed. The bodies of Makin and Michael Quinn were delivered to the surgeons for dissecting while the bodies of Culley and Thomas Quinn were hung in chains at Guyhirn opposite the house where the murder was committed. They were hung by their heads in an iron helmet, which became known as 'Paddy's Nightcap', and can be seen in the Wisbech Museum.

Much of the following records were taken from *Irish Migrant Agricultural Labourers in Nineteenth Century Lincolnshire*, an editorial by Sarah Barber 1982 (SB). Records also show that a surge of Irish labour started coming over to Lincolnshire, mainly to the highland farms rather than the Fens, for harvesting corn from the late eighteenth century. Not everyone took too kindly to the Irish as a report in 1809 from the *Lincoln, Rutland, & Stamford Mercury* reveals, 'A very violent and unprovoked assault was made by drunken English labourers upon six Irishmen and their two wives . . . It appeared to be the object of the English labourers . . . to drive these Irish labourers out of the Country, and deter others from coming into it.'

The start of the steam packet service between Ireland and England in 1822 greatly increased the numbers of Irish who migrated here. At first they came mainly from the west coast of Ireland, walking to Dublin to catch the ferry to Liverpool or Holyhead. From there they would walk around the Pennines over to Lincolnshire with the hope of finding work.

Hostilities from the English did not deter the Irish from coming to work in Lincolnshire and by the 1840s large numbers were arriving, especially around the Stamford area. It was not uncommon for Irish wives to accompany their men folk to work as a team, cutting, binding and stooking the corn. Corn harvest in England would have been two to three months before the Irish potato harvest, enabling the workers to return to lift their own potatoes. The weather played a great part in their chances of earning money to take back home. In 1845, the year of the potato famine in Ireland, many could not find work in Lincolnshire. The *Lincoln, Rutland & Stamford Mercury* reported the same year, 'The result of heavy rains throughout the whole of eastern England left thousands of Irish destitute, waiting by the side of the roads for the harvest to begin, three thousand are said to have passed through Stamford at the end of August that same year.'

The same year the Lincoln Union poor house added another building to its premises to house the Irish who could not find work. There was very little communication between England and Ireland at that time so the migrant workers had to estimate the beginning of harvest and if this was delayed due to bad weather they relied on begging or the hospitality of the Catholic Church. The city of Lincoln, and towns such as Boston and Stamford in particular, were often bulging at the seams when work was not available on the farms.

By 1850 the railway routes were established, chartered trains of cattle wagons loaded with 500 Irishmen were arriving at Lincolnshire towns, but no doubt many still walked the journey to Lincolnshire. Not all the migrants would have been on the breadline in Ireland – some were probably small crofters. On the ferry the fare for 'a decker' would have been around £2 (until the price war between the ferry operators arose, which did reduce the cost later). The train fare to Lincolnshire was 6s and food would be 6s, making a total cost of getting here about 12s.

Wages varied from county to county and even district to district, but an agricultural labourer's wage in England during the late nineteenth century would have been around 12 to 15s a week, with a tied cottage provided. Some workers, such as foremen, horsemen and those tending livestock, would have had other benefits such as milk provided if there were cows on the farm and fuel for heating and cooking. A farm labourer in Ireland at the same time could earn no more than 6s per week.

J. Farrow & Co., Peterborough. Seen here after the Second World War are mostly Irish students who worked at the pea canning factory. Two static viners are in the background.

A gang of five Irish potato pickers on Mr John Richardson's farm at Twenty near Bourne, with farm foreman Jack Massam in the centre of the photograph, 1973.

Potato growing increased at an exceptional rate from the mid-nineteenth century. Rich virgin soils could now be planted with potatoes, which were a staple diet for much of the population during this era. The potato acreage on a farm was determined by the labour force available at harvesting time. For thousands of Irish facing poverty in the nineteenth and twentieth centuries the only alternative to emigration to America was migration to England for work.

The depression in England started to take effect in the period 1870–80 and forced many farmers to change from arable to livestock, reducing their labour requirement. This, together with the use of corn harvesting machinery, also had an adverse effect on the Irish migrant labour force. In 1864 the *Lincoln, Rutland & Stamford Mercury* reported, 'there is now hardly a farmer in this locality of Kirton Lindsey without a reaping machine, having bought the same or hired at the rate of 2s 6d an acre.'

At the time of the potato famine in Ireland in the 1840s it was reputed that 100,000 Irish were in Lincolnshire, which Sarah Barber states in her editorial as probably being an overestimation. From 1894 the numbers reported to have booked on the ferries were a constant 15–16,000. She also states that from 1851 numbers were a staggering 50,000, dropping to 100 in 1981. Whatever the correct numbers, we do know that many thousands did come to Lincolnshire for work on the farms.

Ill feeling towards the Irish workers was still evident in 1916. At Crowle near Scunthorpe, Lincolnshire, the local newspaper reported that:

Employers were wary of employing them because of resentment among their own workers. The Irish were excluded from military service in their own country while English males were compelled to serve in the British Forces during the First World War. While serving in the forces English males were paid 1s per day at the same time Irish men were taking their places in the farms in the Fens being paid between 5s 6d to 7s per day. Feeling ran high in other sectors of society as shopkeepers and innkeepers declined to accommodate the Irish.

Lincolnshire had declined in importance as a corn-growing county and fashions in agricultural economies were changing. It would be to the Fens that the Irish labour now turned for work. Here is a 1915 poem depicting those times:

> They came to the station,
> With their bundles on their back,
> And go down to Lincolnshire,
> And go sleeping in the sack. (SB)

Potatoes were becoming an important crop in the Fens, as were turnips and mangolds for feeding to stock. Mustard seed growing was established, which had to be cut and bound by hand because of its fragile nature and fine seeds.

When mechanised cereal harvesting took over from hand cutting and binding of corn, it was probably the end of Irish female labour on the farms. The male Irish potato pickers during that period preferred filling and emptying their own baskets, whereas local women picked potatoes into baskets that were mostly emptied into carts by men.

Little if any Irish labour came to the Fens for potato picking during the Second World War, but by the 1950s they returned in large numbers. One of them would find suitable men who wished to pick potatoes and gather around him a 'Paddy gang' of between four and six men. Most farms whose main crop was potatoes by then were employing Paddy gangs. They would come over to one of the Fen towns and frequent the pubs where farmers would come to look for potato pickers. Certain pubs were renowned as 'Paddy pubs'. I remember going around some of these pubs with my father looking for Irish labour in the 1950s.

Most farms provided accommodation to house these gangs either in empty cottages that were called 'Paddy kips' or 'Paddy huts'. These huts were usually purpose-built in nearby paddocks or in the farmyards. They were constructed mostly of brick or were ex-Second World War Nissen huts and often only had one room, similar to the army barracks-type accommodation. Inside would be beds or bunks, table and chairs and a calor gas or solid fuel cooker. Toilets were sometimes flush toilets but often soil type outside and washing facilities were simple. They were very basic even in those days, yet the occupants always appeared to be of good humour, especially after returning from the local pubs. They could earn good wages on piece work and most of their earnings were sent home to their families in Ireland.

I remember their needs were simple but I do also remember they never skimped on food: hard work requires calories. They would buy good joints of meat and invariably have flitches of very fat bacon hanging from the ceiling, and they ate copious amounts of it for breakfast and lunch. Beer was a very important part of their diet, keeping the local pubs going with a good trade. It was a common sight to see droves of these men on a Sunday, which was a day of rest, going to a local Catholic church in the towns. Publicans knew this and after Holy Mass they made them very welcome in their 'tap bars'. One local pub down at the bottom of a long fen road was reputed to bring a bath tub into the bar and fill it with beer. When the Irishmen arrived they could fill their large jugs directly from the bath and replenish their pint glasses much quicker than they could get the beer out of the barrel.

By the 1960s potato harvesting machines had been developed and were being used extensively on the black fen and silt soils. Where potatoes were still being grown on the heavier fen soils, and clods were difficult to separate from potatoes, handpicking was still carried out by the Irish labour force into the 1970s.

Entry into the Common Market in 1972 brought a surge in combinable crops prices and greatly increased the profitability of cereal growing. This influenced the heavy land farmers to grow more cereals and oilseed rape and cease growing potatoes. Irish labour had from early days been involved in harvesting cereals and autumn potato picking, but not in horticulture and vegetable growing. As one vegetable grower said, 'I never saw a Paddy cutting cauliflowers.' This was the start of the demise of the Irish labour in the Fens and the birth of a system of farming to be known as 'combinable cropping', where all crops grown on the farms were harvested by a combine harvester. It was the end of another chapter in migrant labour in the Fens.

One of the last recorded gangs of Irish potato pickers in the Fens was probably in 1973 on F. Richardson & Sons' farms at Bourne Fen, Lincolnshire. John Richardson still kept heavy horses for harvesting corn from stook to stack, then using them to harvest potatoes on his 1,400 acres of fenland around Bourne.

EARLY TWENTIETH-CENTURY WAGES

Within a decade after the First World War, farming was once again in a recession and the plight of the farm worker, like many farmers themselves, was in a dire state. Wages had been an issue for many years and while the Labour government had supposedly vowed to better the lot of this sector of agriculture, little was done by them, despite the Agricultural Wages (Regulation) Bill of 1925. A Fenland landlord at the time was the Marquess of Lincolnshire (Lord Carrington) whose estate extended to 25,000 acres including several thousand acres in the Fens. Order of the Day for the Second Reading of the bill read:

> THE MARQUESS OF LINCOLNSHIRE
> My Lords, it will be within your Lordships' recollection that the measure which this small bill seeks to amend was brought in last year in the House of Commons and entrusted to Mr Noel Buxton, the then Minister of Agriculture. He made a very remarkable speech, with which I need not trouble your Lordships, but which showed what a ghastly state of things existed in some of the counties of England, and particularly in the two counties of Norfolk and Suffolk. The only reference that I desire to make to that very remarkable speech is to recall that Mr Buxton stated, on his authority as a minister, that a standard rate of agricultural wages in the county of Norfolk, of which he was one of the representatives –namely, at the rate of 25s per week – an agricultural labourer who had a wife and five children, after he had paid his rent and secured the ordinary necessities and decencies of life, could count on being able to spend only ¼d per meal for himself, his wife and his children.
>
> This statement, so far as I know, has never been challenged, much less contradicted. All of us who have for a good many years done our best with regard to agricultural reform thought at once, and were happy to think, that a Labour party was in power, because we believed that with a Labour party in power such a horrible state of things would not be permitted to last for ten minutes, and that a bill would at once be brought in giving some hope to these unfortunate people and providing them with a certain rise in wages of 5s per week. Some of us hoped that the rise would be 10s a week. But we listened to the speech and listened to it with disappointment. We were told that the policy of the Labour government was to do nothing at that time – to do nothing to help these people over the winter – but to postpone anything that could be done for eight or ten months and to entrust the settlement of these questions to county councils. We were told that the best way to manage the situation was to do it piecemeal, according to the wants and emergencies of the different districts.

Wages nationally at that time ranged from £1 9s to £2 2s per week, while Holland in Lincolnshire was 36s per week. Norfolk was renowned for low wages, which would have reflected on wages in the Norfolk Fens. Wages remained depressed until the late

1930s. A wages ledger belonging to F.H. Smith, a wholesale potato merchant in the Spalding area, for one of their potato gangs records the following weekly wages: a Mr Marriot was paid £2 2s in 1932/3, £2 3s in 1934/5/6/7 rising to £2 6s in 1938/9. The rates being paid varied between £2 2s to £3 10s depending on the individual.

FRUIT GROWING

The fruit growing areas around Wisbech as far back as fruit growing began required an abundance of fruit pickers, as they still do to the present day. Up until this time the acreage of all fruits – soft, cane, bush and top fruits – was determined by the availability of labour during the picking season. The railways changed this by allowing labour easy access to the Fens as well as providing a transport system to convey the produce to the masses in the towns and cities across England.

Entire families came from the East End of London for the fruit picking season. They were housed near the farms or nurseries and even their children were cared for and entertained while the parents and older children worked in the strawberry fields and orchards. University students from Cambridge were also employed to look after the children. For many of the families it was their only chance of going to the countryside while having the opportunity to earn some money in the process. This practice continued even in the years after the Second World War. Travellers were also drawn to this area for work on the fruit farms and still are in a few numbers. Local labour is still available for seasonal work on a limited scale but over the past decade growers have come to rely almost entirely on foreign labour.

Londoners arriving at Wisbech railway station for the fruit picking season in the early twentieth century. *(Lilian Ream collection)*

The Second World War

The Second World War dramatically changed labour requirements in agriculture. Food became a national priority for the second time in a quarter of a century, of which potatoes featured high on the list. Potatoes were the most important crop in the Fens but in those days involved high labour input. With the lack of Irish labour coming for potato picking, local labour became of vital importance, but was insufficient for the needs of the crop.

Agriculture was one of the listed occupations essential to the war effort and labour employed in this sector was exempt from conscription, termed 'a reserved occupation'. To fill the void the government revived the Women's Land Army in 1939. This was not a new concept, only a recreation of the WLA formed in 1917 that served its purpose during the First World War, then disbanded in 1919. Only fit and able-bodied females between the ages of 18 and 40 could join, which in its initial stage attracted 4,484 volunteers in England. Many were attracted by the slogan 'For a healthy and happy job join the Women's Land Army' rather than work in the munitions factories. The job proved rather different to what many had expected, but most soon settled down and enjoyed their work. By 1944, 80,000 women were working on the land across Britain, carrying out work that had been male-dominated before. Most of these women had come from towns and cities across the country and many had never set foot in the countryside before. They carried out all categories of manual work both with livestock and arable farming as well as operating tractors, horses and other skilled jobs on the farms. The Fens had its fair share of WLA girls, as they became known, with many purpose-built camps set up to house them dotted across this area. The force was finally disbanded in 1950 when numbers had dwindled to fewer than 7,000.

Our family had many WLA workers during the Second World War, several of whom became good friends and remained in contact with us, including some who married American servicemen and returned with them to their homeland. The women lived in a purpose-built hostel in Holbeach Drove until the hostel was turned into a POW camp.

The Women's Land Army's Burton Dozen (eleven in the photograph plus the one taking it), who worked for Mr Burton-Thorney standing in front of the thrashing machine. They were billeted at Thorney WLA camp.

Second World
War Italian POWs
riddling potatoes.
*(Lilian Ream
collection)*

German prisoners
at Shouldam Gap
POW camp in 1945.
In the back row,
third from left is
Gotthilf Martin who
stayed in the Fens.

This camp, like many others in the Fens, was an important source of labour for local farms, initially from the WLA, and then Italian followed by German prisoners of war became an important source of farm labour, not only in the Fens but in other parts of the UK. The Fens, however, was the prime source of potatoes in the UK, requiring vast numbers of workers during various stages of the crops' growing, harvesting and grading processes.

The first influx was after the defeat of the Italians in North Africa when 4,700 Italian prisoners were taken. Then came the Invasion of Europe in 1944, after which large numbers of prisoners were taken and sent to Britain, many of whom worked on the land. The main camps in and around the Fens were at Fulney, Spalding, Stamford, Sleaford, Ely, Boston, Trumpington, Friday Bridge, Wisbech, Thorney, Horbling, Feltwell, King's Lynn and Downham Market. Several WLA hostels were turned over too, as the war progressed.

Methods to improve crop husbandry were orchestrated by the War Agricultural Executive Committee (WAEC), which was divided into regions. These committees did exceptional work in promoting better husbandry on the farms, albeit many farmers did not like being told what to do, which no doubt included many Fenmen!

POST-WAR YEARS

In the post-war years, female labour featured highly on the fenland farms. Gangmasters employed women from the towns and villages transporting them to agricultural and horticultural holdings as well as packing stations. As demand for labour grew, their source of workers spread further afield. Many of the large employers of labour had their own buses transporting workers to and from the workplaces. Workers were being collected from distances of 100 miles or more on a daily basis arriving on the farms and packing stations by 7 a.m. and 8 a.m. These workers, being both male and female, often came from areas of high unemployment.

Up until the 1970s the Spalding area was known for its flower industry, both for bulb production and blooms. Forcing of bulbs for blooms during the winter months was a major sector of this industry. All the stages of production required a large workforce year round, and this was mainly made up of women.

Mechanisation in agriculture during the 1960s witnessed the demise of both male and female labour on the farms. During this period large-scale redevelopment and resettlement of deprived areas was taking place across the whole country, funded by the governments of the day. Peterborough and King's Lynn especially were development areas vying to attract firms to move into their catchment development zones. Government grants and incentives encouraged businesses to relocate from major cities and overseas, creating work for the dwindling labour force on the land.

It was a period of depopulation of much of the fen rural areas, the great trek into the towns leaving hundreds of farm cottages empty culminating in their demolition. The road I live on, which was only 6 miles from Spalding and 12 from Peterborough, saw eight cottages, two public houses and our local Methodist chapel demolished, as well as many other buildings nearby. Some large farmhouses were also added to the list. Many of these cottages could have been preserved but local councils forced them to be demolished if they were not modernised, which was not cost-effective for the owners if tenants could not be found. Those cottages that did remain eventually became sought-after properties, often by families who had left the Fens in the 1950s and 1960s and who now wished to return.

With advanced mechanisation in potato growing, the requirement for casual labour declined. With more pre-packing of the potato crop, grading moved off farm to the central pack houses, reducing labour requirements on the farm. However, the other major industry in the Fens, vegetable growing, saw its labour requirements growing. The supermarkets were requiring more processed and prepared food for their stores, which was becoming labour-intensive. Much of this was done by either local labour in the Fens, sourced from nearby towns and cities, or from other parts of the UK. This sector of the food industry grew beyond all expectations as did the demand for workers in the fields and pack houses during the 1980s.

Farming in the Fens was becoming totally geared to the supermarkets and the change in consumer demand. In less than a decade, the demand for labour outstripped the indigenous population once more. As the older women retired from land work, their daughters with higher education qualifications found jobs in the fenland towns and the city of Peterborough. If the Fens were to hold on to its agricultural and horticultural markets, primarily the supermarkets, a large unskilled labour force would have to be found. This was required to pack and process the locally grown produce, flowers and fruit as well as the imported goods brought into the area to maintain continuity of supply for out-of-season demands.

During the post-war years many nationalities, from Europe and North Africa, came to the UK to find work in the fields, nurseries, canning factories and pack houses. Immigrants who came to the UK from Britain's ex-colonies, such as India and the West Indies, did not come to the Fens, preferring the large cities to find work. Fortunately for the Fens, allowing more countries to join the EEC and free movement of labour from the new entrants fulfilled the demand with labour from these satellite countries.

By the twenty-first century, most firms growing, harvesting and packing vegetables in the Fens came to rely totally on foreign labour for their field and packing operations. This does not only apply to casual work, many also rely on foreign labour for their regular staff as well.

The Travelling Community

The travelling community we have in the Fens today come mainly from Irish origins. Many travellers supposedly came over to the UK during the potato famine in the 1840s and previous to that after Cromwell's reign of persecution in Ireland. I have found no evidence of this in the Fens. Many travellers were known as 'Tinkers' for their work mending pots and pans as they travelled. Another name used in the Fens is 'Didicoys', which derives from the Romany language describing inter-bred Romanies with other ethnic people.

Since the nineteenth century the Fens had attracted travellers for seasonal work in many sectors of agriculture and horticulture. They would move around the country in unison with whatever crops required labour for planting, picking, pruning or harvesting. Large-scale growers employed them for planting out various crops such as salads and celery. They knew what time of year farmers and growers required them for working in fields, orchards and fruit fields, often camping on their holdings. The fruit industry, top fruit, bush and tree fruit especially around the Wisbech area, was a centre for work spread over much of the calendar year.

As the demand from the supermarkets for specialist crops grew in the Fens, the travellers ceased to travel for work as much. Their work by that period was mainly in the southern section of the Fens around Wisbech and Ely. This encouraged many to establish themselves in small camps and not take to the road, hence the large numbers in the Cambridge and Norfolk sector of the Fens.

By the 1970s mechanisation of planting salad, celery and other crops gradually reduced the need to employ them. Orchards were grubbed up during this period requiring less casual labour. Growing of cane, top and bush fruit was also on the decline adding to the cut in demand for labour. Many still live in the Fens but few are employed in agriculture and horticulture today.

Gangmasters

One family in the Fens who have witnessed more changes in gang labour requirements, availability, and their changing workload are the Bassett family of Wisbech, gangmasters since 1923. The founder of the business, Fred Bassett, was born in 1895 and served for two years in the First World War in the Cambridgeshire Regiment. After being demobbed in 1917 he worked for a local gangmaster, Joe Barnes of Wisbech.

He was a dedicated worker, which was observed by some of the farmers and growers whose holdings he worked on. So much so that some of them, not entirely satisfied with the gangmaster they worked for, suggested he set up on his own. With work at his finger tips he set up as a gangmaster to supply labour for them in 1923.

The early years of the business were in the depression years of the 1920s and early 1930s. The Fens around Wisbech were heavily involved in fruit, flower and vegetable production reliant on labour during seasonal peaks. Male labour tended to be employed on a regular basis but women made up the majority of casual labour, except for the Irish potato picking gangs. Fred employed almost entirely female labour from in and around

A thrashing machine being operated. The man on the left passes sheaves of corn up to the two men on the top of the machine who cut the sheaves and feed them into the drum. In the foreground is a man bagging the thrashed corn and a man carrying sacks to the store.

Wisbech. Times were difficult for all sectors of society in that era with no unemployment benefits and jobs hard to come by. 'If you could walk you could work,' were words passed down the Bassett family line to Fred's son Joe, born in 1928, now eighty-two years of age.

The effects of life in the trenches in the First World War gradually took their toll on Fred's health and he died prematurely in 1947 aged just fifty-two. His son Joe was then still in his teens (nineteen to be precise) and if the business was to continue he had to pick up the reins. Reins were something he knew well, having ridden in most of the local flat races that were held after the Second World War around the Fens. His great memory was winning the Holbeach Stakes in 1945 on his own horse, Honest Joe. However, his prowess on horseback had to end when he took on the running of his father's gangmasters business.

From this period through to the 1960s saw the zenith of the tulip and daffodil industry, especially around the Spalding area. Joe expanded his labour force to seventy workers, transporting them in his fleet of Ford lorries with canvas-covered rear bodies. Companies such as the Spalding Bulb Company and T.R. Pick Ltd were extensive growers in those years, requiring hundreds of women to work, and this labour was supplied by the Bassetts. All the women were paid on a 'piece work' basis, meaning gangs were paid by the piece, being a set wage for a set amount of work, such as bunches of flowers cropped or an amount of bulbs picked. Thorney was also one of Joe's workplaces with farmers such as Pick, Gee and Hurn. Most of the work here was with potatoes or pulling peas for the fresh market. All their casual labour was sourced from around Wisbech.

The Bassetts had established a name in the Spalding area going back to the 1930s with the growth of the Spalding Bulb Company. The company was set up in the 1930s under the expert management of D. van Konynenburg, a Dutchman who had pioneered the growing of tulips and daffodils in that area as well as the indoor forcing of flowers under glass. H.C.C. Tinsley of Holbeach Hurn, a large farming enterprise as well as food packing concern, was one of the mainstays of the Bassett workforce during the 1960s and '70s, but ceased in 1984–5.

Great changes in labour supplies and work came in the 1970s. Belgium, France Germany, Italy, Luxemburg and Holland were the six original members of the European Common

Market. In 1973 Denmark, Ireland and the UK joined to make a total of nine nations. The UK's entry into the Common Market allowed a free flow of labour into this country, most of them coming from Italy. Many Italian immigrants came to the Peterborough area during the 1950s mainly from the poorer southern part of the country.

Joe and his son Fred were still finding enough local labour for their workforce even after 1973. By the late 1980s with Spain, Portugal and Greece joining the EU, immigrants from these countries were being employed in the area. Portuguese passport holders held by many living in its old African colonies joined the inflow to areas such as the Fens, where unskilled labour was required for the new pack houses and food processing factories springing up to supply the growth in supermarkets. The Bassetts, however, still employed entirely local labour.

The enlargement of the EU in 2004 (A8) bringing another eight countries into the fold was probably the greatest influence on our labour market in the Fens' agricultural history. The travelling communities who had settled in the Wisbech area also made up part of the labour force, being employed mainly on the fruit growing farms and nurseries. However, their employment was curtailed with the influx of Eastern Europeans.

IMMIGRANTS FROM A8 COUNTRIES

Cyprus, Czechoslovakia , Slovenia, Slovakia only made a limited impact on the Fens labour market, with the greatest supply coming from the Baltic countries of Poland, Latvia and Lithuania and only marginal numbers coming from Estonia. The UK, Ireland and Sweden were the only countries in the EU to allow free access to people from these new member states. It would follow that the UK would be their chosen destination for work. With the growth of large agri-businesses supplying packed and processed food and flowers for the ever-increasing demand of the supermarkets, the demand for labour escalated beyond anything witnessed in the Fens' history.

The largest geographical clusters of A8 workers on the Workers Registration Scheme outside London up until 2006 were in and around the Fens. People from these Baltic countries relish working in the outdoors and packing stations handling produce, fruit and flowers, and filled the labour void. With demand for goods from the Fens whatever the weather to fulfil supermarket contracts, workers must work in these demanding conditions, which these people do.

Much of the gang labour in the Fens prior to 2004 was foreign but the Bassetts had managed to maintain a totally English labour force until that year. Local labour dried up, so they changed to foreign labour due to circumstances rather than policy, and it was the end of an era in the labour supply across the entire Fens. The feet working on the fenland soils had changed nationality again. The numbers employed by the Bassetts vary during the seasons but around 25 are permanent staff while seasonal staff can be up to 200 when certain crops are harvested.

In 2006 the Gang Licensing Authority (GLA) was set up, and was reviewed again in 2009, to protect workers from abuse in agriculture, forestry, horticulture, shellfish gathering and food processing and packaging. The GLA operates a licensing scheme for those acting as a 'gangmaster' or 'labour provider'. Licensing standards set out the conditions that must be complied with in order to qualify for and retain a GLA licence. This act also covers accommodation, workers' contracts, health and safety, pay and tax matters, as well as recruitment issues. To many gangmasters across the Fens this was not a problem but served to curtail the growing exploitation of foreign labour from Europe and other parts of the world. The authority has achieved its objectives for the majority of people employed by gangmasters, but human nature being what it is, some will always be exploited by the unscrupulous.

The Bassetts have seen many changes over their eighty-seven years, from English labour to foreign. Cropping and crop handling has changed dramatically but is still labour-intensive. Crops such as strawberries have a longer season, being grown in glass houses, and polythene tunnels which extends the season from April through to August as opposed to a three-week season in the past. Polythene used for covering vegetable, fruit and flower crops has extended the harvesting season creating the demand for labour throughout the year.

It is not only agriculture and horticulture requiring labour from the family. Other non-agricultural industries now rely on seasonal labour, with more peaks and troughs in their marketing and production requirements.

Many of the immigrants are skilled, talented and highly educated in their own fields back in their home countries. They come with one intention: to work and earn money which they do in all weather conditions, and they are a vital part of the fenland agriculture and horticulture industry. There are still gangs, notably around the Boston area, sourced from local labour, but most are foreign. By and large our local labour has moved into the industrial units, or through government handouts have chosen not to work, so this void would not be filled by the indigenous people of the Fens.

Many people from Eastern Europe after a few years, having started educating their children here, will stay, but will easier prospects tempt them from the fields to the factories in years to come? It is not only locally sourced fruit, vegetables and flowers that are handled by these labour forces. Pack houses fill the lines with foreign produce to supplement our seasonal demands.

We have over centuries attracted migrants to drain the Fens and work here, and as long as the incentive is there, and legislation allows it, we pray that is how it will remain. The population is forever growing with more mouths to feed, and the Fens are a vital part of the food chain as are the people who work here. Since the Second World War we have witnessed a change in our fenland labour force of many nationalities who have always filled the demand.

Roberto Divkovic, born in November 1968 in the State of Yugoslavia (in the area which is now Croatia), came to the UK in 1989 as a student while studying for a degree in Economics. It was through the student organisation Concordia that he worked for Maurice Crouch Growers Ltd at Willow Farm, Manea. He worked there for two seasons during 1989/91. A few years after completing his degree in Economics, in 2000 he returned to work for the Crouch family in the role of administration of students who were there through Concordia.

Cropping daffodils on Mary Hay's farm, Holbeach Marsh, in 2009. Nigel Patrick (farms manager) is on the left with gangmaster. The pickers are mostly from Eastern Europe.

The Crouch organisation, which included the 'Merry Mac' brand, was at this period progressing through a period of restructuring of its business. Their farming acreage was increasing to fulfil the demand for their crops but a rationalisation of crops grown was scheduled for review.

The 1990s saw rapid growth of the salad business with the expansion of fast food chains such as McDonalds, Burger King and such like, as well as the rise of 'bag salad' products being produced by food processors and sold by supermarkets generally giving a big increase in the requirement for Iceberg and Cos type lettuce. The demand for hard red and white cabbage in salad bag packs, plus an uptake of convenience foods opened the doors for expansion for this crop.

The directors of Maurice Crouch, having known Roberto for a few years and experienced his working and management capabilities as well as his bi-lingual attributes, decided to give him the chance to provide all their labour requirements. This was a brave choice to employ a non-UK citizen for such a large organisation as Maurice Crouch Growers. He quickly took control of labour management with an eye for efficiency, reliability and excellent labour relations.

In 2005 he set up Roberto Mac, a registered Recruitment Agency to supply premium temporary and permanent staff to local businesses. Through his website, applicants from around the globe apply for work, stating their requirements and qualifications. Each person's credentials, credibility and background are scrutinised to fit the needs of the employers.

Jobs are available for many needs, from the unskilled manual worker to farm machinery operators. Another sector they are providing for is medical and nursing staff in and around the Fens, supervised by Roberto's wife Maria Divkovic. His work also involves providing labour for the Maurice Crouch Growers daffodil operations in Cornwall. Health and safety and risk assessment are all part of the firm's operations, as is a fleet of vehicles that operate ferrying labour to and from work in the UK as well as Europe.

His workforce runs into hundreds of male and female labour of many nationalities, most however come from Latvia, mainly of Russian ethnic background, and Lithuania, all of whom prefer to work outdoors while the Polish mainly prefer to work in the packing and processing units. Slovaks also feature on his wage sheets. On average the percentage of male to female labour is 65 per cent to 35.

To keep up with the demands of a modern training and staff monitoring organisation, larger premises were required, and so Roberto purchased the Old Salvation Army building in March. It is refreshing to see this building turned into high-tech administration offices, yet still retaining its rightful place of historic interest in this Fenland town. In many ways it still reflects its original purpose of helping people, especially the young as they strive for a better life.

There is a common consensus of opinion that if the Eastern Europeans ceased coming to the Fens, our supermarket shelves would go empty. Roberto believes that with the amount of labour looking for work around the world, the void could be filled, if legislation would allow. Labour can be moved as easily, quickly and cheaply as produce which comes to this country. With the world population growing along with the hunger for work, labour will always be available, and our catchment area will spread. We have always had a transient labour force in the Fens and always will.

Macurice Crouch Growers
Labour History During the Past 50 Years

In the late 1960s early '70s they had about thirty full-time staff, including lorry drivers and mechanics along with about thirty or forty 'regular' piece workers for seasonal

Gangs harvesting salad cabbage on Maurice Crouch farms, Manea, in 2010. All this labour is provided by Roberto Mac, March.

work. Half of them were the remaining family and friends of Italians who still lived on the farm, and the other half consisted of local people from the Manea area. By the early 1980s, which included the land in Cornwall, the full-time people fully employed remained about the same. The Italians had gone to the brickyards at Peterborough and the company was using more local gang labour as their own regular piece workers had left or retired. On peak daffodil picking days in the 1980s they could have up to 400 at work and that number is nearer 300 today. The early 1990s saw the arrival of more significant foreign student numbers for seasonal salad work rising from the initial fifty to seventy-five to the peak between 2004 and 2007 of 200 for daily work. The turnover figure of students was probably nearer 500 during a season.

Today the number of full-time staff remains about what it was in the 1960s and '70s, around twenty-five to thirty, but they are more efficient with more productive and better machinery.

From 2007 onwards poorer productivity from students coupled with greater expectation and with increasing immigration restrictions, meant a drastic change was necessary. This was the time they reverted to a specialist labour provider, while at the same time full-time staff numbers were reduced to nearer twenty. Now seasonal, unskilled and skilled staff is provided by Roberto Mac when required.

My Recollections of Farm Labour

My father farmed with his two brothers in and out of the fen. They started farming on their own in 1927 on a rented farm with a farmhouse on it. Over the next twenty-five years they bought or rented several more farms. The fen farms were all equipped with farmhouses and the odd cottage. It was mostly the foreman, stockman or horse man who lived on those farms. The remaining regular labour lived in their own cottages in or near villages with their own small piece of land. Casual labour, both male and female, came in from surrounding towns when required at harvest time and riddling potatoes.

In my early youth, workers were either horsemen who worked the horses that were looked after by waggoners, stockmen who looked after livestock, tractor drivers or simply farm labourers. All would be supervised by the farm foreman. There would also be the yardman who tended the livestock for the house, a pig or two and chickens for eggs and capons. He also worked in the garden and milked the cow twice a day. Piece work was almost always done by the labourers who got little overtime so they could make up their wages this way. Workers' wives and daughters would be employed on a regular basis, as well as their sons if required.

Mechanisation changed our workforce, especially after the Second World War. During my life on a fen farm I have seen vast changes in this field, mainly with the casual labour. During the war, and post-war years WLA, Italian and Germany prisoners worked alongside our regular staff. After the war they were replaced with men returning from the war along with Irish labour. When the Irish ceased coming we relied on local gang labour, both male and female, sometimes with Italians from Peterborough who had settled there in the 1950s. When we changed to a totally combinable crop rotation, very few regular men were kept on. Peak harvesting times were manned by young men from Australia, New Zealand and South Africa who lived on the farm living in vans, only working for about four to five months, vining peas and grain harvesting. I do remember the post-war years when it was not uncommon for men walking the roads to just call in and ask for work and living accommodation. One in particular was called Tom:

> When I was young and the war had past
> men drifted from the cities fast;
> Homes were gone and loved ones died
> with only tears for those who had survived;
> Men roamed the countryside on weary feet
> to look for work, somewhere to live – as well as eat;
> Hearts were broken, lives fuelled by strife
> Demands were few but little did they ask of life.
> Some were good, others not so, father always said
> some he set to work, others just gave bread.
>
> A stranger came on to the farm, name of Tom
> a kindly man who father did take pity on;
> He gave him food and somewhere to rest
> and stayed and became a friend, the very best;
> Living space was scarce, so were bricks and mortar
> father bought a railway carriage, for Tom's-living quarter;
> Some evenings we would go and listen to his rhymes
> sitting round the stove, I did not forget those times;
> Tom moved on, the carriage gone, all with no regret
> he came into our lives and left but I did not forget.

7

Fields Apart

The farmers' footprints on the soil have been steadily disappearing in the fenland fields and those same fields have become stereotyped, having lost their identity. The early practice of strip farming land in a rotational system was not confined by field boundaries. Land not used for arable cropping was often grazed by commoners with commoners' rights and laws of each district. Before the first major drainage schemes in the mid-seventeenth century, most of the area we know today as the Fens was common land. Vast areas of the Fens even in the mid-eighteenth century were not drained or enclosed, with no defined field boundaries. Enclosing was being carried out as late as the early nineteenth century in some areas. Much of the area was void of hedges unlike the neighbouring uplands which had been enclosed for livestock grazing. The Fens' rivers were tidal to the very upper reaches of the fen edge before the seventeenth century.

The areas for pastoral agriculture would have tended to have been divided up by nature rather than man. The creeks leading off the many rivers were natural boundaries in themselves. Some of the early flood banks served as dividing areas of land, as did roddons. These field names either took on the owner's name or often a name relating to their position in an area or sometimes an event which took place on that piece of ground. Tithes also often referred to field names, not acreages, and those areas not enclosed were deemed Tithe-free, although some land under private ownership and agriculture did have tithes into the twentieth century.

Almost every town and village in the Fens has several charities set up by benefactors, some dating back to the seventeenth century, which provided for the poor and other needy causes in their parish. The charitable handouts consisted mostly of small fields or parcels of land, some with cottages often let to the highest tender with the income going to the poor in the parish. The *Gedney Terrier* in 1631 mentions 'Clock land' which was held by the sexton, his duty being to keep the church clock in good repair. The *Fleet Terrier* in 1613 mentions a field called 'Plumb Piece', the rental income providing for the poor in Gedney Parish. Some small fenland villages had up to twenty charities with small fields all let to provide income for the needy.

The enclosures in the late eighteenth and early nineteenth centuries redefined boundaries and ownership resulting in new names for new fields, the size being determined by whether they were worked by horses or manually. Not all fields had a individual identity – sale particulars I have of 1929 state that the Christie Estate sold a parcel of 'land lying in ming', meaning it had 'no defined boundary'.

Changes did come about in the early twentieth century when larger estates were turned over to small-holdings to enable entrants into farming. This happened in many parts of the Fens mainly on the easier working soils suitable for more intensive agriculture and horticulture. Fields of 20 to 30 acres were often reduced to 10- and 20-acre holdings. The county councils were at the forefront of these schemes, as were the Crown, land settlements, Ministry of Agriculture and some private landlords, namely Lord Carrington and the Duke of Bedford.

The period during the Second World War and after also witnessed the enlargement in field size. The advent of new ditching and drainage machinery, together with tractors and larger implements was another factor in the demise of the small fenland fields. Draglines could cut new drains and the spoil was used to fill in the smaller dykes and ditches making larger fields. Machines for laying underground clay pipes improved the field drainage. These large fields created by the amalgamation of small fields took on a single name or number, with the loss of many small field names.

This happened again in the 1970s. As the water table had been constantly lowered during the twentieth century, drainage allowed larger areas of arable land to be under-drained, creating even larger fields. It was very noticeable between the 1960s and the 1980s how many dykes were filled in across the entire Fens to form larger fields. The 1970s and 1980s saw the introduction of large tractors and implements along with combine harvesters from the USA and Canada suitable for these larger fields. Fixed-wing aircraft and helicopters also made crop spraying in these fields more economical and more timely for applying agrochemicals and fertilisers. Entry into the Common Market with higher prices and grants to improve all aspect of farming encouraged prairie farming in the Fens. Government grants were paid to make fields larger, grubbing hedges out and reshaping fields. On one of our family farms, five fields were all amalgamated into one 100-acre field. Five field names were exchanged for one – 'the 100 acre'. We did, however, plant a small wood in the field where previously not one tree was situated in the five fields.

In the 1970s many financial institutions and pension funds bought vast tracts of land in the marsh and Fens. Modern farming methods were used, new buildings were erected, drainage schemes undertaken and conservation work carefully done. Many areas were made good for future farming generations. They also made larger fields, identified by numbers on their computers, also adding to the demise of the field name.

Fields at one time had their own identity, and were archived as a part of our rural history. Names such as Charlie's field, Tom's field and Crow Tree, alongside Lazy acre, have gone, dead and buried only to be reincarnated on a spreadsheet with a number.

Some of the field names on our family's land were Charlie's, No Mans Land, Fallow Hill, Tin Sheds, Long 27, Blackcurrant, Connection, Chase, Lambert, and there were many names relating to their previous owners, Moores, Walkers, Orreys, Farrows and Hensons. Chase as a field name occurs on several farms in areas across the Fens, the word meaning, 'an unenclosed area of private preserve'. Today our Chase field is a 50-acre field but at one time a very small part of that field was a grassy area infested with rabbits, probably left undisturbed as a source of food and pelts, and so the name lives on.

Not all large farming enterprises have lost field identities. On one of the most up-to-date farms in the area, farmed to modern standards, tradition dies hard. Fields have been mapped by satellite but still retain a birthright. A list of some of the fields on Mary Hay's estate leave the reader mystified as to their origin. Many fields were reclaimed from the marsh in 1950 and only bear the name at birth 'Section', while those reclaimed in 1793 and earlier have names of historical interest. Land reclaimed from the marshes in the eighteenth century and earlier was often referred to as 'Newlands'. The following names have been kept for their individuality: Magpie, Crow Marsh, First 20, Second 20, Third 20, Woodstock, Battle Brig, Brick Clamps, Tubs, Coastguards, Marsh 40, Watsons, Flint House, Desert, Mid 30, Bass 40, Bass Entry, Lapwing, Booths, Kip, Near 30, Glass House, Clods, Near 17, Far 17, Old Barn, Far 30, Floor Pits, Horse Gull, Section 1n, 2nd, Section 3 ,4, 5, Nooks, Middle Field, Gull Ground, Newlands High Marsh, Allotments, Blacksmiths, Poly Tunnel, Rowels, Christie and Grange.

Michel Sly has 'Abbey fields' in Thorney village, site of the Thorney abbey grounds. Stephen Grundy has a field called 'Mucky ten' which is actually 10 acres, while Ben Runciman on his land near Weston has a field called 'Beggars Bush field', supposedly where beggars slept under a thorn bush, and another called 'Hills and Holes', a very undulating silt field.

Dick and Mick Lawson handpicking potatoes into bags and a cart in 2010. *(RS)*

Extracts from *The Victoria History of the County of Cambridgeshire and The Isle of Ely Volume 4* from the early nineteenth century mention fields such as Leverington Parish Fields, The Marsh, Spitalfields, Thummins, Margerie's Croft, Farthing field, Outnewlands, Fen Croft, Church Croft, Paps Croft or Hill Croft, Seafield, Ives Dyke field or Dooloe, Park field, Woolcroft or Walcroft, Wrarfield, Snailcroft otherwise Fendyke field, West Meadow, Beaconhoe or Maysfield.

In Gorefield parish nearby, the following fields are mentioned: Ox or Fitton field, Gorefield, Harp, Hart or Harlpley field, Long meadow, Car field, Richmond field, New field, Shire or Shear field, Black lane field. In Parsons Drove and Southea, also in the area, there are fields such as Pock or Pokefield, Remers field, Popefield, Canon field, Old Eaufield, Elbow field, Flain field, North and south Inham and Parson Drove Fen.

These are but a few names linking us with our parochial history. On many farms and holdings I have found fields when added to an existing farm have taken on the name of their previous owner, and yet none bear the name of the present owner or his ancestors. It seems to imply that to have a field named after your family, it has to change ownership. And the soul remains in that soil.

OS NUMBERING

Field numbers are no longer shown on Ordnance Survey maps. However, they were shown on Ordnance Survey county series maps which were introduced in the 1840s which would have been added to the maps as and when each area was surveyed. Parcels of land were originally numbered consecutively throughout each parish with the numbers printed on the map, and until 1884 the areas, together with the land use of each parcel, were recorded in separate Parish Area Books. This system was discontinued after 1884 and areas, to three decimal places of an acre, were printed on the maps below the parcel numbers. Some changes were introduced to accompany the revision of the series beginning in 1891 but unaltered parcels retained their existing numbers and areas. New parcels formed by partition of an existing parcel also retained the parent number, but suffixes a, b, c, etc. were added to distinguish new divisions. Where older detail had been altered, some parcel numbers were cancelled, and some parishes, the boundaries of which had been changed, were renumbered. After the First World War, when revision was confined to areas where the need was most urgent, it was no longer possible – in as much as one map might be revised but not its neighbour – to show the total areas of parcels broken by sheet edges. From 1922 areas of parcels were measured to the map edge only and this policy was carried on into the National Grid series up until the point where Ordnance Survey ceased to show parcel numbers on its maps in the mid-1990s. Today all our fields are mapped digitally using GPS and supplied to farmers through DEFRA.

WHAT'S IN A FIELD NAME?

Fenland agriculture today has some of the largest grower organisations in the UK alongside a lingering number of small, predominantly family farmers and growers, each with their relevant markets for their products. The large growers have expanded on the back of the growth of supermarkets, processors and marketing companies. These organisations need large volumes of farm-grown products to sustain their requirements which could only be supplied by highly efficient large-scale growers. To grow these products, large acreages of land are required which growers have gathered around them through purchases, contract farm agreements and farm business tenancies.

The two latter methods have encouraged farmers to release their land for such agreements, and generally, these farms have been anything from holdings of 100 acres to upwards of 1,000 acres. They have gone down this route for many reasons, such as not having viable holdings to reinvest in modern systems which many specialised crops need, family commitments or holdings which are insufficient size to support the family's needs.

The marketplace for crops grown in the Fens has changed since the Second World War as the general public have moved to the supermarkets, resulting in the demise of the fresh fruit and vegetable shops and town markets, who were supplied by the wholesale markets who themselves have declined. The catering market has grown, noticeably for fast food and other eating establishments, supplied by a reduced number of wholesalers either direct with growers or the remaining wholesale markets left in the UK. Where large supplies are required for these outlets it is the larger growers' marketplace. Auctions still survive in the Fens at Wisbech and Spalding which has been the lifeline for the small grower to sell his produce.

It has by and large, with some exceptions, become a two-tier grower system: large growers supplying bulk orders of varying quality to large buyers, and smaller growers supplying niche markets, such as shops, stallholders, farm shops and auctions. The reason the small grower has survived is because his marketplace has survived whereas

Lettuce planting machine on Shropshire farms near Ely on black fen in 2009. There are sixteen people feeding plant modules into the machine. Those walking behind fill in any which have been missed. *(RS)*

Lettuce planting machine viewed from the front in 2009. The machine is self-steering following the mark left by a previous cultivator which operates by a Global Positioning Satellite device. *(RS)*

the medium-size growers have gone. They are in themselves a unique breed of people who can survive by hard work, a bonding with their soils and they accept nature's gifts and misgivings. The Goliaths of the farming world are of the same ilk, but have more of the fenland adventurer spirit in them, both drawn from the soils. The following two examples of the modern farming area.

LINCOLNSHIRE FIELD PRODUCTS, WOOL HALL FARM, WYKEHAM, SPALDING

The business has its roots in the van Geest family stretching back to 1929, but following a decision of the family to divest itself of its interest in the original business in 1997, changes were made. Three members of the then senior management team, Robin Hancox, Aubrey Day and Martin Tate, successfully completed a buyout of the business. In the run up to the buyout, the business was under the direction of Robin Hancox, managing director from 1980. Since then emphasis has gradually shifted from a more traditional arable farming company to one with particular focus on the growing, procurement, marketing and distribution of high-value vegetable and root crops.

There are many aspects of this farming enterprise which set it apart from most others in the Fens. Its scale of operations is the most striking part; in 2010 they will farm 16,000 acres almost entirely on Grade 1 silt. The soil is land reclaimed from the sea over many periods of history, from the Roman era right through to the eighteenth century. For this reason much of the land is not in prairie-size fields, some are part of smaller holdings divided by dykes, medieval sea banks and winding creeks.

In the past this area of the Fens was made up of smaller farms and holdings unlike the Great Levels which were drained in the seventeenth century and shaped and sized for large-scale farming in the twentieth century. Now long, straight droves and drains with names like forty foot, twenty foot, sixteen foot or even ten foot drains divide up the farms.

The farmed area is not in a single block but all within close proximity of the hub of the main operations unit at Wool Hall. The present Wool Hall was probably on the site of an older house, and as its name implies, it was a wool merchant's house when the marshes were grazed with sheep back in the seventeenth and eighteenth centuries. Early medieval settlements and reclamations intermingle around the farming enterprise, overseen by the proximity of fenland churches nestled in the villages and towns of Spalding and Holbeach. I have not set foot in all their fields but it would surprise me if a church cannot be seen from almost every one.

The second aspect which sets it apart from so many farming operations in the Fens is it has only recent history, some thirty years in all, as Robin Hancox pointed out, 'if you are going to write about history, we have little.'

Lincolnshire Field Products is not a family farm, as we know the meaning, as it has no family history. It is a farming enterprise conceived in the late twentieth century to grow, prepare and distribute food to the shopper via the supermarkets. It could be said that LFP was conceived for the era of the supermarkets and twenty-first century food production. The area farmed has been gathered together through the changing face of managing land for farming. The advent of contract farming and farm business tenancies (FBT) has changed the farming industry to suit the modern agri-farm business such as this one. It has also enabled existing farmers who wish to retire or cease farming to hold on to their land for personal reasons, yet reap the rewards from agriculture.

Robin maintains that by letting their land to LFP on a contract farming agreement, whether it be 20 acres or 200 acres the owner himself becomes part of the organisation supplying supermarkets. He can still retain the ownership of his land along with reaping

Cauliflower cutting rig near Spalding in 2010, with Robin Hancox speaking to the gang. *(RS)*

the benefits of the economy of scale. Many such smaller farmers are doing this and finding jobs or diversifying into other enterprises.

Robin has large acreages of land on FBTs mainly where farmers wish to rent him land for cash crops. Many of the crops LFP grow are specialist crops which some farmers cannot grow themselves either because of their complexity, the lack of marketing outlets or the investment required in specialised equipment. LFP's strength is its marketing expertise with guaranteed outlets and prices, which comes with its scale of operations.

Robin was born in Northamptonshire and went to the local agricultural college at Moulton, where he completed the NDA. He then had sixteen months working for A.H. Worth & Co. on Holbeach Marsh in 1976/77, after which he went on to do a range of jobs around the country broadening his experience. Eventually, in 1980, Robin returned to Spalding to work for the van Geest family.

The enterprise we see is mechanised farming technology at its cutting edge, integrated agronomy of the highest conceivable status. The cropping for 2010 consists of the following acreages: Broccoli 3,200, Cauliflower 3,300, Brussels sprouts 500, Cabbage 1,180, Potatoes 1,400, Beans 360, Oilseed Rape 304, Wheat 3,131, Peas 900 and Sugar Beet 1,800 with a total 16,075 acres farmed. The sugar beet crop has seen the most noticeable increase of all the crops during recent years.

In 2009/10, turnover was £93 million, with the business employing 260 full-time staff and up to 300 seasonal workers. Produce is chilled and packed in all manner of packages in the most modern buildings in the land, with the consumer in mind.

FreshLinc is a sister company of LFP based in Spalding. It specialises in refrigerated distribution and storage, not only for their own goods but for other firms around the UK. At any one time it may have 190 refrigerated vehicles on the move, day and night, to deliver produce to its destinations.

The business has grown through demand for its products mainly for two supermarket chains, ASDA and Tesco. Between them they take 80 per cent of what LFP produce. Large acreages, large equipment and guaranteed outlets are fine, but large crops of quality produce are what must be produced to fill the order books, this all coming from the soil.

BRIAN (SONNY) TEMPLEMAN, SMALL-HOLDER, SURFLEET, SPALDING

In the early twentieth century during the allotment and small-holding movement, a small-holding was deemed to be between 1 and 50 acres. A century later, with 40 acres to his name, Sonny can still claim that identity. He was born in 1933, the son of Alfred Templeman who worked for a prominent local farmer, Captian Edward Proctor, who farmed across the River Welland from where Sonny lives today. Alfred always said he could never have had a finer mentor than Captain Proctor and he branched out on his own by renting a 1-acre allotment, which was owned by the Surfleet Charities. Later he made the allotment up to a 3-acre and eventually rented some land. The Surfleet Charities still has land today which is rented out to local small-holders and farmers; the organisation is made up of three charities, the Joseph Burton Charity 1735, Samuel Elsdale 1810, and the Lady Fraiser Charity 1764, although some of their deeds date back to 1640.

Sonny served his National Service in 1951 then returned to work for his father on his small-holding in Surfleet until 1968 when his father retired. The family had bought land and also rented some close by and when his father retired, Sonny had to buy his father's share of the business which was the working capital of the small-holding. To his dismay 1968 turned out on be one of the worst years in farming since the 1930s due to terrible storms and flooding in July, followed by adverse weather throughout the remainder of that year. However, he survived and carried on to build his holding up to 40 acres, the last purchase coming in 1970 when he bought the 10 acres he had been renting. This is in an area where farmland is very expensive and sought-after by neighbouring farmers if it comes on to the market.

All the work carried out on the holding is done by himself, his wife Carol, son Martyn and daughter Clare, most of it involving manual labour. His cropping consists of 10 acres of cauliflowers which is programmed to be cut throughout the year, some covered with polythene wraps for early harvesting, enabling him to have something to sell all the year round. Other crops are a ½ acre of sprouting broccoli, ½ an acre of brussels sprouts and 5 acres of sugar beet, lifted by contractor. He grows 4 acres of potatoes, which are all picked by hand, and 3 acres of daffodils, some of which are forced in a glass house for winter cropping, while some are cropped in the fields as blooms in the spring. A few tulips are grown also to be forced in the winter under glass and some are sold as bulbs.

Not all of his land is cropped by the family; he rents some to other growers for vegetable growing in an area where premium prices are paid for this type of land, all being good Grade 1 soils. He sells his vegetables and potatoes to a local farm shop, a stall holder and shops in the neighbouring town of Bourne as well as Lincoln. The Spalding Auction is also a vital market for all his produce and flowers, some of which also go to the New Covent Garden in London.

The success of his survival in an area where many growers are now farming many thousands of acres has many facets. The quality of his soil, which he owns, is a key factor enabling him to grow almost any crop he wishes. He also has regular cash flow, continuity and hard work and identifies the correct markets for everything he grows. He would call it nothing more than 'common sense' and does not need anyone to advise him how to run his holding. It is the small grower who has also helped the small trader

The Templeman family, Surfleet, cutting cauliflowers in 2010. *(RS)*

Sonny Templeman, his wife, son and daughter cropping forced daffodils in their glass house, Surfleet, in 2010.

to survive in a market dominated by multi-nationals and large growers. The large growers may not compete directly, but sometimes do dump surpluses into the auctions which can undermine the small growers' market. There is a loyalty, however, between the small grower and his buyers which if it remains, both will survive.

Another example of the small grower is Willie Chapell, age fifty-five, who farms 50 acres in 3rd Drove, Gosberton Fen. His cropping consists of 10 acres of cabbages and cauliflowers, 7 acres of potatoes, 1 acre of onions and the remainder is spring wheat and winter barley. Apart from the wheat and barley, all his produce is sold through the Spalding Bulb and Produce Auction. Almost all the work is carried out by himself with occasional help from some neighbours.

8

Crop Spraying

Machinery, more than any other factor in farming, changed the face of British agriculture after the Second World War. The combine, potato, and sugar beet harvesters cover the land only once in a year, as does the plough, baler, drill and several other farm implements. The agricultural sprayer, however, can traverse a farm many times during the seasons covering many thousands of acres. Crops will require several applications of agrochemicals during their growing seasons, depending on their type and variety. The weather and disease susceptibility will be a deciding factor in the chemicals applied.

The first sprayers most arable farmers had were the horse-drawn dusters used for applying copper sulphate to control blight in potatoes. These machines were used into the 1940s after which Bordeaux Mixture and several other copper-containing fungicides were introduced. Machines were mounted on tractors with booms which sprayed the liquids onto the crops.

Chafers were a company who supplied agrochemicals and built machines to apply them. One such machine was to apply Dithane, a copper-based powder for controlling potato blight used in the late 1950s. Spraying cereals, peas, beans, sugar beet and almost every crop grown in the Fens became the norm. Trailed, mounted and self-propelled machines with ever increasing boom width were used on every holding.

The years between the end of the war and the 1960s saw the fastest development in insecticides, pesticides and fungicides, together with machines to apply them. The need to spray these materials regularly through the crops at optimum times caught the eyes of the aviators.

AERIAL SPRAYING

The USA were the pioneers of aerial spraying for pest control. Using mainly powders, the operation became known as 'Crop Dusting'. They began in the 1920s with ex-First World War aircraft. The application of fertilisers from the air was pioneered in New Zealand during the 1940s using De Havilland Tiger Moths. However, the real breakthrough came after the Second World War when thousands of Piper Cubs used for pilot training were sold at extremely low prices in the USA. Many pilots from the war were keen to carry on flying and took up aerial crop spraying and dusting.

In the UK the application of agricultural chemicals from the air was first used after the First World War using the de Havilland Tiger Moth, but only on a very limited scale. Little if any was done during the Second World War, after which the Tiger Moth was soon joined by the Auster.

East Anglia, more especially the Fens, was the perfect platform for this revolutionary agricultural operation. Large, flat fields, few woods and trees and a predominately arable cropping were the ag-pilots' dream. Their biggest hazards were electric cables on pylons and poles, but these pilots soon learnt to combat the challenge by keeping low to the ground and flying underneath them.

A Tiger Moth crop spraying in the 1930s.

G-APIG DH 82A Gipsy Moth de Havilland 1F Moth, built in 1940, was owned by Westwick Distributors Ltd of Norfolk and is still flying in France. Here it is on display at a farmers' open day in the early 1960s.

One of the pioneers of fixed-wing ag crop spraying in the Fens was Cliff Annis from Boston. Cliff joined the RAF in 1941 and trained in the American Air Force scheme. After returning to England he trained on various aircraft, eventually flying the Lancaster. Indeed, the Lancaster aircraft he captained was the first bomber in the war to achieve fifty bombing operations, but not with the same crew. For Cliff and his crew it was their ninth operation in that aircraft. Their tenth operation was to Nuremberg three days later. They took off from Elsham airfield in the Lincolnshire Wolds, not so far from his native town of Boston. The flight had started with the loss of power on one engine, delaying their arrival to the target. On their run in to bomb, a night fighter attacked them and put one port engine out of action. Undeterred they carried on and dropped their bombload on target. Soon after, they were attacked again by a second night fighter which hit the starboard engines setting the plane on fire, after which Cliff gave the order to 'abandon the aircraft'. Cliff survived but was very badly injured and was repatriated through the Red Cross back to England. The other two surviving crew members spent the rest of the war in a prisoner of war camp.

Hardy & Collins rest room in the 1960s.

Surviving his wounds and being disabled in one arm, his love of flying returned. Cliff was initially involved in training, pleasure and charter flying from Boston and Skegness airfields after the war. Skegness Air Taxi Services was one of the pioneers of aerial spraying in the UK. A subsidiary company, Aerial Spraying Contractors, was set up to look after this aspect of the work trading until 1956, which became Lincs Aerial Spraying Company operating out of Boston airfield. Auster Aiglets were used initially with wind-driven pumps for supplying spray to the booms. To maintain work for men and aircraft out of the UK, season work was taken on in North Africa. Three of these aircraft were ferried to North Africa in the autumn of 1950 where they were used for spraying cotton in the Sudan, the contract lasting until 1954. These were very hazardous conditions for men and aircraft. One aircraft was lost during spraying operations while another ditched in the Mediterranean when returning to England – both pilot and passenger were lost. At home three Aiglets crashed, two while spraying and one due to an accident near Skegness.

A Westwick Air Services Piper Pawnee being loaded with Nitrogen fertiliser from a Bedford lorry in 1974.

A Hardy & Collins of Boston Piper Pawnee crop spraying. Note the marker boy at the end of field in white overalls.

The Piper Pawnee PA-25-150 was one of the first purpose-built aircraft for crop dusting and spraying, developed in the USA and launched in 1956. The original aircraft were powered by 150hp Lycoming engines. As demand increased for larger payloads, the PA-25-235 was introduced in 1968, with an upgraded 375hp Lycoming engine enabling them to carry payloads of 1,200lbs.

Cliff purchased his first Pawnee in 1963 and by 1968 was operating four PA-25-235 aircraft. The work by that time included spraying a wide range of crops, applying fertilisers and slug bait as well as seed broadcasting of wheat during wet autumns. They were covering around 90,000 acres of farmland per year.

The maintenance of aircraft and operations was run from Boston airfield. When operating, each aircraft was backed up by a fleet of vans, tankers and personnel which moved from landing strip to landing strip spread across their operational area. When spreading fertiliser, an aircraft could land, refill and take off within 25 seconds. The whole business was extremely accurate, efficient and in today's jargon had a 'low carbon footprint'. Each farmer they did work for were provided with maps of his farm with

Grumman Ag Cat crop spraying peas on Terrington Marsh in about 1970. The sea can be seen in the background over the sea bank.

field numbers so LASC could identify the field when asked to have work done. We on our farms still use their maps. At the peak of their operations in the 1980s, LASC were operating five Pawnees which serviced in excess of 200,000 acres in Lincolnshire and the bordering counties. The business ceased in the 1980s.

It was fate which brought Cliff Annis into aerial ag spraying; had it not been for the Second World War his feet may never have left fen soil. His love of flying all types of aircraft, from trainers to bombers, from vintage to the ag planes, fulfilled his dreams and made him a leader in his field. His skills were also passed on to others, instructing newcomers to the skills of flying, or just talking aeroplanes!

HELICOPTERS

G. & G.S. Neal of Holbeach who started in 1948/9 were early pioneers of aerial ag work. Originally they were land drainage contractors, having machines for laying tile drains under the soil. Gordon and Geoff both had private pilot licences and owned an Auster which they kept on the family farm. Previous to this Geoff had been in the RAF for his National Service and was in Berlin during the Berlin Airlift in 1949.

Although familiar with fixed-wing aircraft, they decided their chosen field in the aerial ag crop spraying business would be with helicopters, which were a relatively new form of aviation transport. However, they were not the first to use helicopters. Dr Ripper was one of the innovators in this field and through his work the first aerial spraying was developed by Pest Control Ltd, part of the Fisons Fertiliser Company based at Bourn airfield, Cambridgeshire. In 1945 Fisons were using the Sikorsky R-4 helicopter and three years later the S-51. Later the business became Fisons Farmwork.

In 1946 Westland UK were licensed to build helicopters in this country and in 1950 they brought out the S-55 which was used by likes of British European Airways and Christian Salvesen Ltd. In 1958 Neals chartered a Westland S-55 from BEA and commenced aerial spraying in the Fens, swapping to an S-55 from Salvesen during the season. So successful were they, to fulfil demand they chartered two S-55s in 1960 from Salvesen. Salvesen at that time had been using the S-55 for whale hunting off the Falkland Islands.

Westland S-55 series 3 owned by BEA operating for G. & G.S. Neal, Holbeach, on the Crown Estate at Sutton Bridge in 1958. Note the ex-army Bedford 15cwt truck used as a spray tanker.

A Djinn 1058/CDL spraying potatoes, operating for G. & G.S. Neal Holbeach in the 1960s. It was the only tip jet helicopter to go into large-scale mass production, and was made in France.

Smaller machines were really needed for ag crop spraying and so Neals changed to the Djinn helicopter, which they hired from Sepatom in France. This was a revolutionary machine at the forefront of jet technology whereby the rotors were powered by jets at the tips of the rotors and not through a gearbox. This technique was partly acquired from the Germans who were working on it during the war. Neals, with others, formed a consortium 'British Executive Air Services' and acquired five Djinn helicopters in the mid-1960s, one of which suffered a fatal crash in 1968. By 1970 they had changed to the American Hiller 12 Es hired from Sloane Helicopters of Sywell.

Entry into the Common Market changed the face of farming as large increases in the acreages of cereals and oilseed rape coincided with demands for more intensive crops. The frozen foods industry required more products to freeze, especially peas. The potato industry was expanding into frozen chips, pre-packs, crisps and other convenience foods. This reflected on the demand for aerial ag contractors. One of the most hectic seasons for Neals was 1978 when they hired three Bell 47 Gs from Benair, a Hughes 300 and at one point a Hiller 12 E.

To cope with the five helicopters, three teams were needed to service them consisting of one supervisor, one fieldsman, four markers, one tanker driver and one aircraft fitter. From the 1960s through to the 1980s demand for their service work grew, but weather was still the main player in affecting demand. Wet seasons made it sometimes impossible to operate ground machines, not only for spraying insecticides and fungicides but applying fertilisers and slug pellets. Wet seasons were also responsible for potato blight and other fungal diseases in most crops. Fixed-wing and helicopters really came into their own being able to apply these products as and when required. Notable bad years for potato blight were, 1958, 1968, 1978, 1982 and 1987, which although bad for farmers were good for the aerial ag contractors.

Flying under pylons, electric cables at very low altitude requires extreme skill from the pilots, many of whom came from as far away as New Zealand. The 1970s was a decade of several crashes, but none fatal. Neals, together with Shell Chemicals, developed weed control in water courses (Bi-Flon) applying the chemical by helicopter, a very efficient operation in its day. By 1989 they were down to one Hiller and sold the entire business in 1995. For the last decade of Neals' involvement in aerial spraying, Gordon's son John

G. & G.S. Neal's Hiller G-HILR having just completed loading Nitrogen for top dressing in 1980. The hopper was driven by a hydraulic motor.

Hiller operating under electric wires spraying potatoes in the 1970s.

was operations manager, a section of the business he ran with a passion and was sad to see it end.

The aerial ag spraying operators came and went in about forty years in the UK, and there were several reasons for their decline. Ground sprayers were becoming larger with wider spray booms, and they used low ground pressure tyres enabling them to operate in wetter ground conditions without damaging the soil. These machines were expensive but farms were becoming larger and investing in larger, more sophisticated machinery. Although applying agrochemicals and fertilisers from the air was still very cost-effective at that time, the public and the government were far more concerned about the environment than they were during the immediate post-war years. The fenland towns and villages were eating into the farmland rapidly by people who knew nothing about how food was produced. Legislation greatly increased the number of notifications and consents the operators had to give to various individuals and bodies before spraying crops. This increased restrictions on where they could spray.

G. & G.S. Neal Holbeach. Two of G & G.S. Neal's early ground sprayers with twin saddle tanks alongsidean ex-Army 4x4 lorry used as tanker for filling sprayers in the 1950s.

Supermarkets also began to pressurise suppliers into not using aerial spraying for routine applications. As the cost of supporting product labels began to rise, chemical manufacturers began to query the cost of keeping aerial clearance on a label, particularly the lower-used products.

There were arguments for and against fixed-wing versus helicopters. Operating costs were higher for helicopters against fixed-wing. The helicopters could land in almost any field for loading the materials they were applying while the fixed-wing needed a runway maybe some distance away from their target fields. Getting the sprays down into the crop whatever the argument was due to the skill of the pilots of either machines. Pilot hours were restricted to an 8-hour day over 28 days and if he wanted more he had to have a medical. The limit was actually on how many hours they could fly in any 28 days. Work rates varied, affected by the weather, and wind speed being the most crucial factor. Crops were sprayed working upwind to allow drift back onto the sprayed area and to make it easier for the marker boys. These were often college students who stood in the crop at each end of the fields as markers for the aircraft moving along on each spray run. Many of the pilots could fly within inches of their targets, and it was not unheard of for pilots to leave their aircraft tyre marks on the roofs of vehicles used by the marker team. Many other daredevil stories were told about these pilots, no doubt some were exaggerated.

Our family used both fixed-wing and helicopters operators on our farms. In the early days it was very exciting to witness the skills of the pilots and we enjoyed watching them perform under wires and pylons. The whole business was very professionally run, in the office and in the field. It was extremely efficient in helping Britain replenish its food stores which had been dangerously low from 1940 until the 1960s. In today's climate it may seem not to have been an environmentally friendly method of producing food, but I believe the aerial ag operators should be looked back on with gratitude for their contribution at that time.

Other major operators in the Fens at that time were:
Fixed-wing: Hardy & Collins, Boston. Westwick Distributers, Norfolk. Miller Aviation, Wickenby.

A Housham self-propelled sprayer on Michael Sly's farm, Postland, with 36-metre spray booms in 2010. It is spraying spring beans down tram lines in the crop.

Helicopters: Fisons Farmwork Services of Holbeach. G. & G.S. Neal of Holbeach. Garners of Sutton Bridge.

Ground sprayers range from tractor-mounted, trailed and self-propelled machines with all sizes of boom widths and varying capacities. In the 1960 and '70s boom widths were on average no more than 12 metres while now most are in of excess of 24. Fertilisers are applied with sprayers as well as chemicals. Tramlines revolutionised the efficiency of spraying, eliminating under- and overlapping in the crops. Air conditioning and air filtration in the cabs also came in the 1970s making the operators' job safer, healthier and more user-friendly. Highly accurate systems on the machines for applying sprays have added to cost-saving and beneficial use of agrochemicals. Training of operators and recording data have turned the spray operator into a highly-skilled agricultural worker on a par with any of his industrial cousins. With agrochemicals being the highest input of the farmer and grower budgets, there is little margin for error. Legislation, traceability and economics have made this operation the most important on any farm.

1. The author in the Fens.

2. The Sly family on the Fleet Coy sluice on the South Holland Main Drain, which was demolished soon after this photograph was taken in December 2009. Left to right are George, Malcolm, George, Derek, Edward, James, Rex and Michael. The children are William, Lucy and Sam. *(RS)*

3. Mary Hay in a field of daffodils on her farm in Holbeach Marsh in 2010. *(RS)*

4. Tulips in flower with forcing houses in the background belonging to W. Bateman of Holbeach in the 1950s.

5. A fen blow. *(Photograph courtesy of Neville Bailey)*

6. A farm building on black fen soil showing how the surrounding land surface has been lost to shrinkage and erosion. *(RS)*

7. Dick and Mick Lawson planting potatoes with a two-row 1960s Johnson potato planter at Shepeau Stow in 2009. *(RS)*

8. Two automatic planters on LFP farms near Spalding in 2009 which plant potatoes and apply fertiliser. Robin Hancox is second on the right. *(RS)*

9. Planting potatoes on black fen soil in Holme Fen, once part of Whittlesey Mere. This soil is prone to blowing. *(RS)*

10. Ploughing on fen skirt land in Thorney Fen in 2010. *(RS)*

11. Peter Cannon and his son Paul feeding cattle in the yards prior to being turned out on the Washes for summer grazing. *(RS)*

12. Peter Cannon's bull with a herd of cattle on Whittlesey Washes. *(RS)*

13. Two MF 760 combine harvesters working at sundown in the 1980s. *(RS)*

14. An aerial view of the John Saul intake in progress in 1972. Note the amount of green marsh left on the seaward side of the new bank.

15. Sonny Templeman of Surfleet boxing forced daffodils from the greenhouse bound for New Covent Garden in London. *(RS)*

16. Sixty-five-year-old John Claxton with a Cat Challenger. He has worked at Wingland all his life, firstly for F.K. Bass and then, since 1964, for Proctor Bros – and he's still working! *(RS)*

17. Oilseed rape in flower in the Fens. *(RS)*

18. A storm sweeping across the Fens in the springtime.

19. Edward Newling of Gorefield inspecting Braeburn apples prior to picking in the autumn. *(RS)*

20. Beehives in the spring in Edward Newling's orchards at Gorefield for pollination. *(RS)*

21. Peas for harvest being put on poles for drying in the field on John Richardson's farm at Twenty in 1976. Left to right are J. Brennan, Bill Yarwell, Dick Beddoes and John Richardson.

22. Left to right in the 1980s are Len Tabiner, Fred Quantrill and Alice Sly (my mother). Father and mother always took tea, cake and beer for the men at harvest time. Both men worked for the family all their working lives and in the garden after retiring. *(RS)*

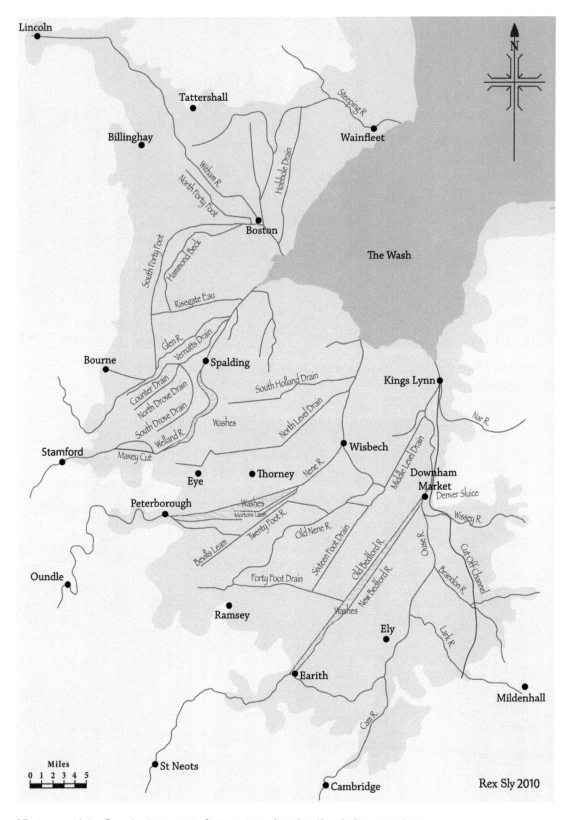

23. A map of the Fens in the twenty-first century showing the drainage system.

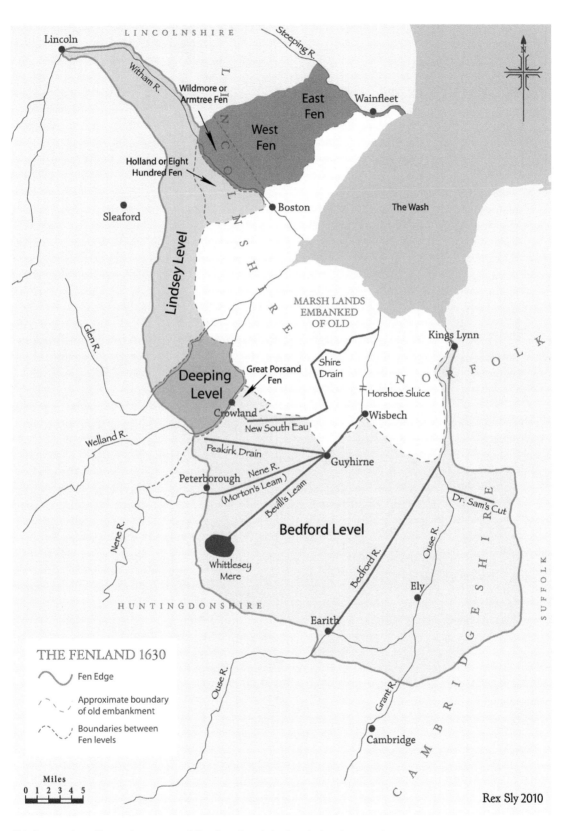

24. A seventeenth-century map of the four levels to be drained.

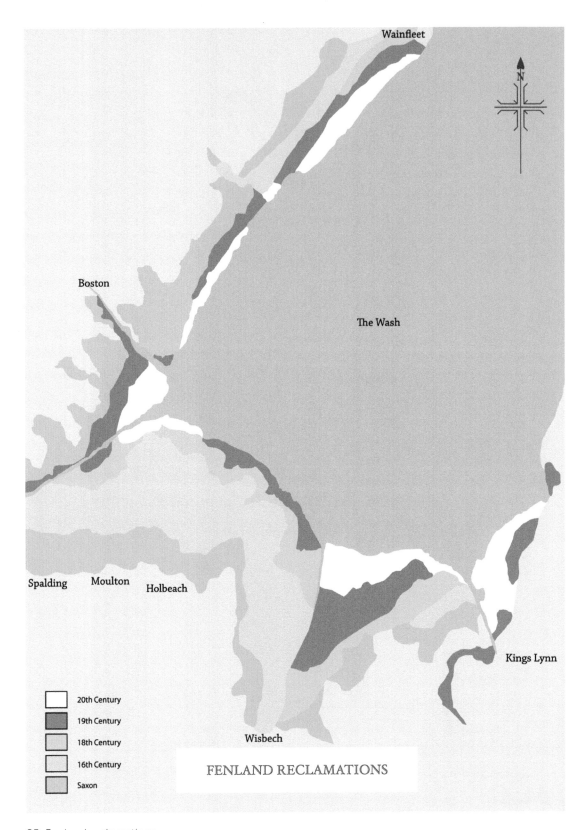

25. Fenland reclamations.

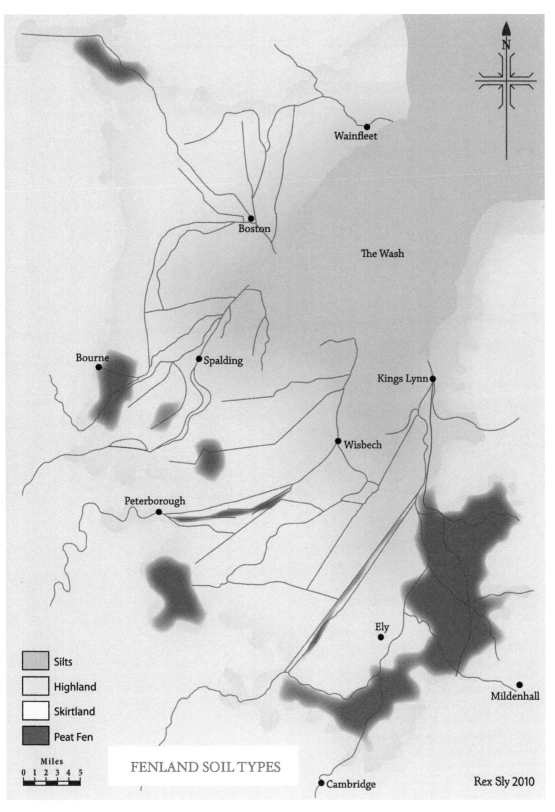

Wainfleet

Boston

The Wash

Bourne ● ● Spalding

Kings Lynn ●

Wisbech

Peterborough

Ely

Mildenhall

Silts

Highland

Skirtland

Peat Fen

Miles
0 1 2 3 4 5

FENLAND SOIL TYPES

Rex Sly 2010

Cambridge

26. Fenland soil types today.

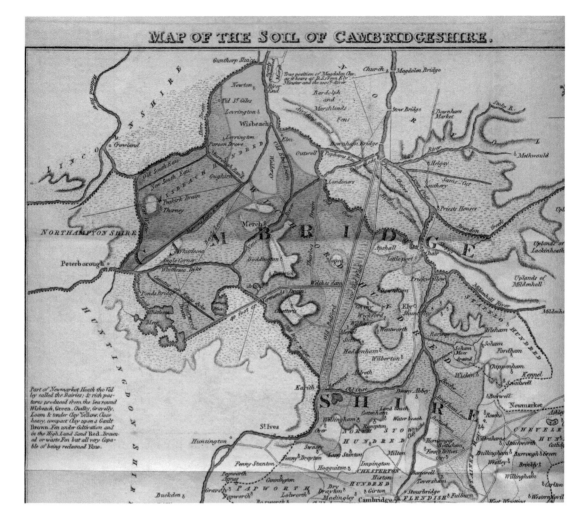

27. 1878 map of soil types in the Cambridgeshire Fens.

Part of Newmarket Heath the Val
ley called the Dairies; & rich pastures produced from the Sea round
Wisbeach, Green. Chalky, Gravelly,
Loam & tender Clay Yellow. Close
heavy, compact Clay upon a Gault
Brown. Fen under Cultivation and
in the High Land Sand Red. Drown-
ed or waste Fen but all very Capable of being reclaimed Blue.

28. Legend for above map.

Fenland Cropping

FLOWERS

The growing of narcissi in the Fens began on a commercial scale in the late nineteenth century. The tulip was introduced in 1905 by W.T. Ware, an entrepreneurial nurseryman at Wisbech, and the flower arrived at Spalding two years later, courtesy of Samuel Culpin. It was the birth of an industry which would label Spalding and the surrounding area as 'Tulip land'. Yellow-flowered bulbs are called daffodils and the red cupped variety are mostly call by their generic name narcissi – yet they are the same family.

The First World War delayed the expansion of this new industry but growers multiplied mainly around the Spalding and Wisbech area by the 1920s. Pioneering names at that time were Dick Welband, Sam Culpin, J.T. White and Oscar D'Alcorn. Kelly's *Directory of Lincolnshire* for 1922, lists 36 bulb-growers most of whom were within a 10-mile radius of Spalding, the ideal land for tulips and daffodils being silt soils found around these growing areas. The silt soils must be of moisture-retaining quality, not too light yet not too heavy so that it restricts the development of the tulip bulb during the growing season. Daffodils require similar soils but will tolerate slightly heavier soils than tulips.

Rapid expansion of the two crops took place until the Second World War, which resulted in the acreage being cut back with growers only allowed to maintain a nucleus of basic stocks. Prior to the First World War both daffodils and tulips were being grown for multiplication and being forced under glass during the winter months for blooms. Rail links enabled flowers to be on sale in the cities around the country within 24 hours of being cropped. The tulip growers grew mainly for the bulb itself, where the heads were taken off by hand while in flower to encourage the bulb to multiply and stop disease occurring from the decomposing petals on the soil. Bulbs were then lifted, cleaned and either sold as bulbs to wholesalers or used to force under glass. The tulip bulbs have little, if any value after forcing – some, though, were fed to

Two women cropping daffodils working for Fred Bassett, gangmaster, Wisbech, 1950s. In those days daffodils were cropped with more bloom than today.

Flowers being delivered by various modes of transport to Cowbit railway station, ready to be taken to various destinations around Britain, 1940s.

cattle. Waste from forcing did become a problem and much of it was carted many miles from source to fill pits and dykes. This was carried out without realising the effects of spreading soil-borne diseases around the Fens and its consequences, as had happened with potato-infected eelworm soil and waste.

Daffodils were mostly left in the soil for two years before lifting, whereas tulips needed to be lifted, cleaned and the bulb split up then returned to the soil annually. Daffodils were usually cropped for their blooms in the spring with the bulbs being marketed for forcing or sold as bulbs to wholesalers. Cropping in the field has little detrimental effect to the bulb.

By 1933 the number of bulb growers in the Spalding and Wisbech area had increased to 150, growing 2,500 acres. These were mainly local families with one Dutchman among their ranks, Cornelius Slooten of Spalding, whose family is still there today. The Dutch would become dominant players in the industry in the post-war years. Some of the local families in the industry were: F.H. Bowser, Eric Casson, G. Bateman, O.A. Taylor, Len and Horace Braybrooks, Mathew Dearnley, Fred D'Alcorn and Alf Cunnington.

The industry expanded rapidly after the Second World War with many Dutchmen taking the lead in the growing industry, but more especially in the forcing of tulips and daffodils. The brothers John and Leonard van Geest and Dominicus van Konynenburg (Spalding Bulb Company) became the two largest bulb growers and forcers of flowers in the country. The Geest organisation grew 750 acres of tulips and 1,500 acres of daffodils at its peak. Other Dutch families to make their mark were Moreman, Nell, Lindhout and Bushman.

By the 1950s and early 1960s road transport had superseded rail, with local firms setting up to cater for this new industry. These crops were extremely labour-intensive, especially tulips with labour having to be brought in from far and wide, but they did prolong the life of the small growers after the Second World War. Many families with a few acres and a glass house made their living from tulips and daffodils, which they would not have done without these crops. It also kept many women in work in the surrounding villages without travelling into towns. Spalding was the prime area for tulips and daffodils, while Wisbech also had a thriving bulb industry as well growers supplying rose bushes to garden centres. Mr Kooreman of H. Prins Ltd was a major supplier of bulbs to Woolworths, as well as a grower of roses.

The 1960s was the peak of the flower industry in the Spalding area, with 8,000+ acres of daffodils, 3,000 acres of tulips and 130 acres of glass for forcing these two crops. The daffodil industry was slowly being mechanised using existing potato equipment and stores; as a consequence the acreage of this crop was expanding, not only in the Fens but anywhere they could be grown in the country. Clean, disease-free land of sufficient quality was becoming difficult to find in the Fens for tulips. The industry had grown to the point where fifty growers, with Len Braybrooks of Cowbit as chairman, formed the Spalding Bulb Exchange in 1961, meeting at the White Hart in Spalding. However, interest waned and the exchange was finally wound up in 1968.

Members of the Bulb Exchange, Spalding, in the 1960s. *(Courtesy of Lincs Free Press)*

Outdoor daffodils being cropped for Mothering Sunday on Cowbit Wash by East European labour in 2009.

Tulip sales waned due to rising growing costs, especially labour and with more homes using central heating which reduced the life of the forced tulip and favoured the ever-growing market of pot plants. The Dutch were able to grow better quality bulbs on their lighter soils and force at a lower cost of production. As a result the tulip industry in the Fens declined and most growers ceased growing tulips by 1990. The odd one like Chapple of Spalding lingered on until 2000. Fresh flowers flown in from every quarter of the globe all the year sold by supermarkets also contributed to the demise of our tulip industry.

The acreage of tulips today in the Fens is insignificant, but the daffodil crop has been the reverse, with our growers being able to outplay the Dutch with our production costs being lower. This is mainly due to the economies of scale with growers handling them with their existing potato equipment, mechanising the crop during the growing, harvesting and planting operations. The flower crop in the spring has become a valuable source of income, especially for peak times such as Mothering Sunday and Easter. Blooms are cropped and put into cold stores ready for timely marketing, spreading the labour workload. There is at this time an ample supply of labour for this work, many coming from Eastern Europe to the Fens just for the daffodil cropping season.

Most tulips in the UK are imported now, but our daffodil growers are major exporters to Holland, USA, Germany and Scandinavia, where they are used for forcing. Ironically, many daffodils exported to Holland find their way back to the UK, as forced flowers and pre-packed bulbs to be sold in garden centres.

The industry has left a legacy in so far as we have major flower importers and packers around the Fens who developed from the growth of these home-produced crops. Other legacies of those once-great days are the gardens at Springfields and the Spalding Flower Parade, once the showpiece of the industry. In 1950 an annual Tulip Queen was chosen when thousands flocked to this area to tour around the tulip fields, which developed into the Tulip Parade in Spalding in 1959. The tulip industry came here because of our soils, lasted less than a century, but left its mark in fen history, and is still the emblem of Spalding.

Daffodils being loaded into a trailer by T. Charlton and his sons at Easter 2006. His family have farmed at Moulton for five generations. Left to right: Henry, Tom (senior) and Thomas. (RS)

FRUIT

Fruit growing had developed within a 10-mile radius around Wisbech from the early nineteenth century. The soil lends itself to all types of fruit growing due to its fertility, moisture retention and workability. The town of Wisbech was at that period a thriving port. Timber was brought in for making boxes, punnets and fruit trays and fruit could be shipped to other ports around the UK. The advent of the railways in the mid- to late nineteenth century made the industry even more important to the local economy and the nation. Wisbech was on the Midland railway line which joined the LNER line at Peterborough, one of the main lines running north to south. Fruit could now be sent anywhere in the UK.

Specially ventilated box wagons were built with well-sprung undercarriages and were filled with containers purpose-built to carry fruit over long distances. These trains became known as the 'Fruit Trains'. They were attached to express trains delivering fruit to the major centres of population overnight, to be on offer at the wholesale markets next morning.

Extra labour was required as growers expanded their acreage of fruit which could now come from the major cities such as London. Horse manure was brought from London by rail for use on the farms. Around Wisbech, canals, local railway lines and a tram system were used to collect fruit from the surrounding farms. A fast fruit train went through our farm on the Northern line, every evening, at the same time, heading north. If we were working in the field downwind of it in the summer months, fields we used it as our time piece to finish work, just by the smell of fruit. The next fast train was the fish train, if we smelled that we had missed supper! Fruit was categorised into the following:

  Apples, pears and plums as top fruit.
  Raspberries, loganberries as cane fruit
  Strawberries as soft fruit
  Gooseberries, redcurrants and blackcurrants as bush fruit

Around this industry the processing, packing, canning and freezing industries blossomed. Many more businesses earned their living from the fruit growing industry, such as haulage, freight, box-making and many aspects of packaging material, label printing and other suppliers. The auction houses sprang to life to sell the growers' wares to customers from around the UK.

Strawberry pickers of all ages, probably from London, in Wisbech in the 1930s.

WISBECH & DISTRICT FRUIT GROWERS ASSOCIATION (WDFGA)

Reading through the association's year books from the 1930s it is evident that the industry has secured markets across most of the UK. The firms advertising in their handbook extend from Scotland to the south coast of England and as far west as Belfast. There were fruit salesmen and merchants from almost every major city in the country. One merchant's advert depicted a line of factories' smoking chimneys along with a long row of houses also with smoking chimneys. The caption read, 'Where Chimneys Smoke there is a ready buyer for your Produce'.

Fruit and flowers were sent to the markets and sold by salesmen or brokers, the grower receiving a price after the merchant had deducted his commission for selling. The market price was dependant on quality, availability and the fickleness of the buyers as well as the general public's whims and fancies. A few of these merchants and salesmen had been established since the early nineteenth century but most after the advent of the rail links. Some included in their adverts 'Account sales and Cheques daily or as desired'.

Chemical and fertiliser merchants and manufacturers were also prominent in the early yearbooks as were articles and reports on their applications and results. The myth in the minds of many of the general public today is that chemicals and artificial fertilisers are a modern conception. However, it is noticeable how many pests and diseases were a problem to the fruit grower in the early part of the nineteenth century. Many of the articles refer to research being carried out not only in the Wisbech area but across the whole country. Chemicals and fungicides such as tar oil, mixed oil and petroleum, lime sulphur, arsenate of lead, Bordeaux Mixture and nicotine were all used in the fruit industry. Paris Green and metaldehyde was used to control slugs in various soft fruit crops. Some of the organic fertilisers advertised in their yearbooks are interesting recollections for today's readers: Fine Dried Blood, Fish Guano, Meat and Bone Meal, Hoof and Horn Meal, Dissolved Hide Meal, Bone Meal, Fish and Blood Compound, Charcoal, Sulphate of Ammonia, Nitro Chalk, Nitrate of Soda, Potash Nitrate, Sulphate of Magnesia, Finely Powered and Dried Sulphate of Iron, etc.

Most of the insecticides and fungicides were applied by hand-operated sprayers although some of the larger growers were also using horse-drawn sprayers with hand-operated lances. What is evident is the constant programme of sprays and washes being used, probably more than today. Unfortunately, few growers keep a detailed account and records of their spraying programmes.

Even after the Second World War there were many chemicals used in the fruit growing industry which were unknowingly injurious to humans and wildlife. Possibly the demand to increase food production during the Second World War had allowed this to be overlooked, seeing there was no alternative. Some, however, were being scrutinised as having detrimental effects on humans and wildlife by the late 1950s. Note from Wisbech and District Fruit Growers Association, 1959:

> In view of the findings at an inquest on the death of a nurseryman that death was due to poisoning by an organic mercury compound, the sub-committee devoted a great deal of time on the question of the use of these compounds, and especially ethyl mercury phosphate. Other chemicals being looked at were Aldrin, CMU, Dieldrin, Dipterex, Malathion, Maleic Hydrazide and Metasystox.

For almost a decade after the Second World War fruit growers were applying more fungicides and insecticides that any other sector of the farming and growing industry. Systematically these chemicals were withdrawn from the market over the next two decades, being deemed dangerous. Today this would not happen, as better testing methods are used before a product is available to the growers, and scrutinised by the government departments and virtually all crops grown in the UK for human consumption are subjected to traceability. All growers of any size supplying supermarkets and processors especially have to, by law, keep meticulous records of every operation they perform when applying chemicals to their crops.

The apple industry in the early twentieth century was under pressure for foreign imports. In 1904, 3,028,000cwt of apples were imported and by 1935 the tonnage had increased to 7,272,000cwt. The main reasons were the demand for the lower acidity dessert apple from around the world and filling the gap during our late storage season. The long-term storage of apples and pears left a gap in the market for the British grower which was filled by the foreign apple. Cold storage was in its infancy in the early 1930s. An advertisement by J. & E. Hall of Dartford in Kent from the WDFGA in 1933 reads:

Cold storage made it possible for the British grower to compete with the severe competition he has to face from imported fruit, especially out of season. Cold storage is being successfully used by enterprising British growers to meet this competition. Of about thirty cold stores for home-grown fruit erected in this country we have been responsible for twenty-seven, including several equipped to provide the new and highly successful system of Gas Storage for apples.

There was a swing to newer lower-acidity varieties being grown in the UK and an increase in acreage. Imported apples had been duty free until 1931 when 58.5 per cent were of foreign origin, after which duty was levied on foreign imports but not Empire imports. The home industry was also establishing fruit research stations at Long Ashton (for cider research) and a fruit research station at East Malling. Migrant labour from London and other large cities came during fruit picking season as did the travellers and Cambridge undergraduates. Overseas students from all parts of the world came through the student organisation Concordia to Friday Bridge Camp. The trend has changed somewhat over the years from UK-sourced gang labour to Eastern European workers. Much of the growing and handling of fruit has become mechanised but work such as pruning and picking is still labour-intensive, especially where quality fruit is required.

Apparently, in 1933, 600 women pickers were brought chiefly from London and Sheffield through the Employment Exchange service of the Ministry of Labour. Harold C. Selby (*Strawberries in the Wisbech District: the first seventy years*, 1977), says, 'Visiting pickers totalled three thousand or more . . . at the pinnacle of strawberry growing era. In 1977 the total Wisbech District crop has changed little at around 8,000 tons.' Much of the strawberry crop grown in the area was used for processing and when large imports started to arrive from Poland, the gradual decline of this crop began.

There are references in the WDFGA yearbooks on agricultural piece work rates for the Wisbech district covering crops such as sugar beet, peas, potatoes, thatching, corn harvesting and seeds harvesting, but there is no mention of any rates for fruit workers.

A FAMILY OF FRUIT GROWERS

The Newling family have been growing fruit on the land at Gorefield for three generations. The present generation is Edward, born in 1957, whose grandfather Michael (1899–2001) bought Richmond Hall along with 60 acres of land in 1926. The estate, which had been owned by the Peckover family of Wisbech, consisted of 500 acres, much of which was bought by Cambridge County Council and split into small-holdings. Michael had two sons, John, born in 1928, and David, born in 1929.

Fruit growing has been the family's passion for three generations. It began when Edward's grandfather Michael bought Richmond Hall and planted his first orchard

Female workers employed by the Newling family fruit growers of Gorefield, Wisbech, 1940s.

there. Both his sons John and David farmed their own farms from leaving school. The pruning shears were picked up by his son David who, after buying Bank farm, Fitton End, in 1957, established orchards there. Since that time more acres have been added to their fifty-three varieties of fruit enterprise with an ongoing replanting programme. There was a restraint on the planting of new orchards during the Second World War, 'fruit deemed not an essential food commodity.'

Michael retired in 1976 and his land at Gorefield was split between his sons John, who took on the arable land, and David who took on the fruit growing side. David's son Edward manages the fruit farm today which is devoted totally to growing top fruit on 72ha. What arable land he does have is farmed by his cousin David. Like many farmer and grower operations today they have become specialists in the growing, harvesting, storing and marketing of their crops, albeit it with far fewer crops today than they had years ago. After the last war their cropping consisted of outdoor flowers such as peonies, daffodils, and chrysanthemums; glasshouse crops like tomatoes, peppers, mushrooms and lettuce; bush fruit such as gooseberries and blackcurrants; root crops such as potatoes and sugar beet; cereals such as wheat and barley; soft fruit such as strawberries and top fruit such as apples, plums and pears – seventeen crops in total. There was also on the farm a pig breeding and fattening herd from which manure was used for the fruit crops. In those days gooseberries and blackcurrants were planted between the trees and strawberries on the pathways running lengthways between the trees. Trees were pruned to a great height and ladders used by the pickers to pick fruit, although now trees are pruned to allow all fruit to be picked without ladders.

Strawberry pickers at the Newling family farm in the 1930s. *(RS)*

Edward Newling inspecting Bramley apples in one of his stores before being moved to his pack house in 2009. *(RS)*

When Edward came on to the farm in 1977, flowers, mushrooms, lettuce and tomatoes had been dropped from the cropping list. Gradually other crops were dropped until the present policy of all top fruit was taken, consisting of apples, pears and plums. Edward was reluctant to follow their policy but he decided keep the family tradition of fruit growing alive when many neighbouring growers had grubbed up their orchards and reverted to arable farming. He studied fruit growing at Hadlow College in Kent to familiarise himself with the most up-to-date techniques.

In the late 1970s there were approximately 1,200 commercial growers in the UK – now there are around 300. Many factors have brought about this decline; economies of scale, retirement with no succession, grants for grubbing orchards and the continued downward pressure on prices. The import of fruit and juices from around the world has also added to this decline. There is a future in fruit, but only if you became a specialist in that field. With less crop, timeliness and scale of production can be fine-tuned to sustain a profitable business.

In the pack house where Bramley apples are being graded and packed for Sainsbury's supermarkets in 2009. *(RS)*

All their fruit in the past was sold through the wholesale markets around the UK. However, with the gradual dominance of the supermarkets Edward began supplying apples to Sainsbury's in 1984 through his membership of a local co-operative, moving to the Society of Growers & Top Fruit Ltd in 2004. Today 50 per cent of his crop goes to them for Sainsbury's, while 20 per cent goes to the wholesale markets, 10 per cent goes to be made into juice, another 10 per cent goes for peeling and the remainder goes for waste.

Vast changes to the business have been carried out over the past twenty-five years. After the war in 1945 a programme of replanting began. Some Conference pear seedlings had been in their nursery beds for 5 years which were planted by Women's Land Army girls at 100 trees per acre and are still in production today. Planting densities have changed from those days, as in the 1960s apples were planted at between 35 to 150 trees to the acre, today that figure has increased to 700 per acre. Another change has been from storing fruit in bushel boxes, each one handled manually, to using pallet boxes handled with fork-lift trucks. Spraying with hand lances and horse-drawn sprayers has been replaced with tractor-drawn sprayers applying fine mist sprays. The labour for picking is still hand labour and will remain that way in the foreseeable future.

They labour itself has changed many times. Since the period of rail transport, entire families came from the East End of London by the thousands on working holidays to pick top, bush and soft fruit when in season. This system carried on through the Second World War when travellers and local gang labour gradually replaced them. As the EU enlarged in the 1980s workers from Eastern Europe took their places. Edward uses mainly local labour in his pack house and Eastern Europeans for picking

Fruit being picked into boxes to be stored in controlled environment stores until packed and graded, 2009. *(RS)*

and pruning who are all sourced by him through word of mouth. Many have now been with him for several years.

His aim is to grow the perfect apple, not in quantity but quality. The cost of production today is so high that he cannot afford to grow, harvest, store and pack second quality fruit. This process begins in the autumn and spring with the pruning of trees. Then, when the fruit has formed on the trees, they remove the king apples, blemishes and to reduce the numbers of fruit. Pests and diseases have always been a major problem with fruit growing so a comprehensive spray programme is carried out, under a very strict supervision, which is recorded for traceability. Picking is carried out with a velvet touch, as is handling into and out of store. They are literally treated like eggs to avoid any damage whatsoever, with inferior apples left in the field. Second grade and damaged apples go for peeling and juice, which are lower priced and not what Edward is trying to achieve. Of the 72ha down to fruit trees, 40ha are Conference pears, 20ha Bramley apples, 10ha dessert apples such as Braeburn and Kanzi and 2ha of Victoria plums.

The fruit are picked into pallet boxes and taken into store where they are stored and monitored monthly to determine eating quality until they are required for marketing. The stores are temperature-controlled using 1 percent oxygen and 5 per cent CO_2 to maintain their keeping and eating quality. They are taken out of store and the pallet boxes gently submerged in water causing the fruit to float on water which transports them to the packing and grading plant. There they are inspected by skilled staff and the grade one apples are wrapped like jewels. A satisfying sight for a grower like Edward is to see the best apples a grower can produce presented in such a manner for the

Picking Braeburn apples using special containers to minimise damage in 2009. Most of Edward's pickers are from Eastern Europe and come over on a regular basis. *(RS)*

consumer. This is all achieved after a year's hard and sometimes frustrating work. Frost and cold winds are a constant worry in the spring. The wind and hail storms can destroy a crop which is days from harvesting or decimate the quality before your very eyes, as can pests and diseases. Edward recalls in recent years when hail storms once thrashed the fruit within a week of harvesting. The fruit was only fit for pulp and went to a cider maker in Ireland.

The dominant players in the fruit industry are the supermarkets, who are encouraging growers to replace old orchards with new ones, but seem reluctant to pay for the necessary investment by the growers. Records in this book, however, show that poor prices have been muted since the beginning of the last century. Many growers fell by the wayside but a few growers like Edward's family have survived by adapting to the marketplace.

With all the state of the art equipment for growing, handling, storing and grading and agronomy, Edward has not lost his basic knowledge of husbandry. Every spring when each variety comes into flower he hires in bees for pollination at 2 hives per ha, still a traditionalist at heart!

This is a 'family concern', by the true meaning of the words; Edward's wife Angela is involved in the running of the business and when riding her horses, sees the orchards with interest during every season. Edward's father John, at eighty-two years of age, still sets foot on the farm in both Sunday best and overalls.

MUSTARD

Mustard, it is said, was introduced to Britain by the Romans and as we know they populated many areas in and around the Fens. The earliest reference I have found of mustard in the Fens was in a poem concerning a Water Party on Whittlesey Mere in 1669. The poem was originally in Latin but translated into English in *Fenland Notes and Queries* and recalls a great feast of extravagant fare and revelry, 'the Jaded appetite again is plied, with sugar, mustard, and cayenne besides.'

William Henry Wheeler (1868) also mentions mustard as coming almost entirely from the fen districts of England and from Holland. A mustard market was held at Wisbech during the months of October and November which the agents of the principle manufactures attended. Mustard was, at that time, grown largely on the peat lands in the East Fen and Deeping Fen, and also on the alluvial lands. Brown and white mustard was grown, the former considered to be of the better quality as it produces most and fetches the highest price, but it requires the best land. White mustard stands better in bad weather and tends to shed less with harvesting as late as September whereas the brown requires harvesting earlier. He states, 'the growth of mustard was first commenced in the Bedford Level in the early part of the nineteenth century and gradually spread to Lincolnshire.'

Arthur Young makes no mention of the crop in Lincolnshire in his report of 1799. The Revd W. Gooch, however, in his 1811 Agriculture of the County of Cambridgeshire lists mustard as grown in the neighbourhood of Wisbech and Outwell. He explains the husbandry involved in growing and harvesting, as well as the yield and prices expected. He also notes, 'it always remains in the land, and many, I was told, it is an impoverisher.' He also says, 'a second crop is obtained from what shells of the first.'

Gooch quotes prices of 56s per quarter, while Wheeler quotes 20s in 1848, 15s in 1858 and 10s to 12s in 1868. Wheeler also states how lucrative this crop was in the case of a Lincolnshire farmer who took a wagon load of mustard to Wisbech Mustard Market and sold it for 50s a quarter, realising £500 for the load. In Kelly's 1902 *Directory of Cambridgeshire*, it lists mustard being grown extensively in the area.

During the twentieth century and even today, mustard is synonymous with Colman's of Norwich, who started manufacturing mustard in 1814, but in the early nineteenth century there were many other firms involved in the mustard industry. Wherrys of Bourne were having mustard grown for them on contract by local farmers for Keen Robinson & Co, a major mustard manufacture, later to become part of Colman's in 1903. The Keen family were one of the founders of the industry having been established in 1742 as mustard manufacturers. Perhaps this is where the phrase 'as keen as mustard' originated from?

In the late nineteenth century Mr and Mrs Joseph Farrow began making mushroom ketchup in Holbeach but as the business grew they moved to larger premises at Boston. Here they extended their range to include dried peas packed in packets and manufactured mustard. In 1902 they moved to Peterborough and built a new factory close to the city's rail links, sourcing their peas and mustard from the surrounding Fens. By this time J. Farrow & Co. were a sizeable company in both mustard and peas. During the early 1900s J. & J. Colman Ltd were buying most of the mustard manufacturers across the country and in 1912 bought Farrows. Farrow & Co. business carried on during the First World War without much change but in the post-war years, salad cream and sauces were added to its range of products. In 1929 the mustard manufacturing was closed at Peterborough and moved to the Colman's factory at Norwich; at the same time the canning of peas was introduced at Peterborough.

The established growers of mustard for Farrow & Co. remained with Colman's and mustard became an important crop for many fenland farmers. Wisbech Corn Exchange remained an important market for mustard with regular visits from the Colman family to purchase mustard seed. The firm also had a stand in Peterborough Corn Exchange.

After the Second World War mustard was grown on a contract between farmer and Colman's, negotiated annually. With Colman's attending weekly corn exchanges in the pre-Second World War years, purchases of seed were probably made from growers or other merchants on the trading floor, indicating that the crop may have been grown without contract. Seed varieties were mostly named after villages in East Anglia, several of which were fen villages. Brown mustard varieties named after fenland villages were Newton and Sutton, while Tilney, Thorney and Gedney were white mustard varieties.

Some varieties of mustard grew to the height of a man or even higher, especially on the rich fen soils. It was sown in the spring, about late March/early April time, and kept clean by hand-hoeing or horse-drawn hoes. Harvesting prior to the mechanical reaper, around the turn of the nineteenth century was carried out by hand labour. The seed was cut with a sickle and tied into sheaves which were left in the field to dry. Later it was put into a stack and thrashed by hand, and in more recent times by the mechanical thrashing machine. This process remained the sole method of extracting the seed from the stems until after the Second World War and in some cases until the late 1950s on fen farms. The crop did have a tendency to shed easily and great care was required when handling the crop prior to thrashing. Large sheets of hessian laid on the floor or ground were used in the process to catch the shed seeds. Even to this day, the crop requires very few chemicals and fertilisers, yet still has a great yield potential.

The crop was widely grown across the Fens for Colman's of Norwich until the post-war days. By then, Colman's were importing large volumes of mustard seed from Canada, blending it with English seed, yet still selling it as 'English Mustard'. The crop gradually declined in the 1960s and 1970s when farmers were unable to grow the crop as cheaply as the Canadians. Oil seed rape came in the 1970s and became a viable and lucrative break crop, especially for the heavy land farmers. Mustard was sown in the spring and often subjected to dry conditions and not always easy to establish, whereas oil seed rape was sown in the autumn. The number of growers for Colman's declined rapidly through to the present day, with little of Colman's seed now sourced in the UK. In the Fens today, the crop is still grown by only a handful of growers which accounts for most of the crop produced in the UK. These remaining growers who were loyal

Mustard for Colman's of Norwich being steerage-hoed for weed control in 2009. Left to right are Neil Clifton, Bill Burton and his son Tom Burton on their farm at Ponders Bridge. The family have been growing mustard for four generations.

to Colman's posed the question to the firm, 'If we cease growing for you will you still be able to sell your mustard as "English Mustard"?' The firm had also done no plant breeding of new varieties for many decades and the farm-saved seed they were using was beginning to deteriorate in quality. Since that meeting with Colman's, eleven existing growers formed a co-operative, the English Mustard Growers Co. Ltd, to preserve the production of this crop. They also took on the task, with the Fenland firm of Elsoms of Spalding, a plant breeding programme, to further its future as an economic part of their crop rotation. With this type of expertise involved and pressure to revive the English mustard crop, it would be a feather in the cap of those remaining growers to see its revival in the Fens, where it belongs.

The acreage of the crop grown for Colman's today is about 1,500 acres by thirteen growers, twelve of whom are in the Fens area. The crop is still drilled in the spring and harvested in late harvest, dried and stored on the farm before being despatched to Colman's Carrow Works in Norwich. The 'Fen Tigers' have stirred from their lair, and a handful of them may even have saved this once-great fenland crop and preserved the name of 'English Mustard'.

ORGANIC FARMING

John and Jane Edwards farm 600 acres in the South Lincolnshire Fens, their two holdings being of similar acreages. New Farm, Wrangle, is Grade 1 silt and Ostlers Farm, Stickney, is Grade 2 fen edge. John's father and grandfather both farmed small-holdings in the Wrangle locality, the turning point coming in 1956 when John's father, Frank Arthur, bought the 44-acre New Farm, Wrangle, for £7,500. New Farm had been part of Wrangle manor which in 1946 was sold by Herbert Sharpe to the Masters of the Fellows & Scholars of Corpus Christi College, Cambridge, which they sold in several lots in 1956. Since that date the family have added to this acreage, both hiring and buying small lots which came on to the market, one lot being a 16-acre field in 1966 for £4,300.

Their land around Wrangle was farmed on a typical crop rotation for that area consisting of vegetables, potatoes and sugar beet together with cereals for break crops. A pig enterprise was also on the farm until 1974 which it closed due to poor returns, which eventually turned into losses.

Their lives changed dramatically in 1984 when Jane (née Ostler) inherited a 270-acre farm known as Ostlers Farm in Stickney Fen. The farm had been owned by the Ostler family, corn merchants of Boston, and let to a dairy farmer by the name of Horace Ward. It lies on the fen edge (Wallasea 2), a very different soil to which John had been used to. The farm at that time was unique, all grass and had never seen artificial fertilisers or agrochemicals, truly sustainable agriculture. It was a time in farming when the word organics was becoming fashionable, but only in certain quarters, and certainly not around Wrangle.

When John was asked what drove him into organic farming, his reply was unhesitating and blunt: 'the wife'. Jane thought it the perfect opportunity to venture into organic farming with advice from the Soil Association. On Ostlers Farm they ploughed some of the grass up and drilled 53 acres of Axona spring wheat, the crop yielded well and the harvest was sold to the Maud Foster Mill in Boston for organic flour, encouraging them to plough up more of the grassland to convert it to arable.

The first attempt with an arable crop using no artificial fertiliser on the holding was very successful but wheat needs to be followed by a break crop. The break crop chosen was spring beans, which put nitrogen back into the soil, which were milled and fed to their beef cattle; wheat for organic flour could then be grown after the bean crop.

This was also thought to be the perfect environment to convert their beef cattle enterprise to 'organic beef', which they did with advice from the Soil Association. To sustain the cattle in the winter months some of the arable land was returned to a 2-year grass ley used for making silage, as well as having clover in them which is a natural provider of nitrogen for the following arable crop of cereals.

More land was added to this holding by renting 60 acres adjoining the farm. The farm was entered under the 'Countryside Steward Scheme' where land is taken out of crop production and set aside for environmental purposes with grass and wild flower mixtures. No potatoes and vegetables are grown on this farm, only on land at Wrangle.

Cattle were not new to John and Jane – they had cattle prior to the pigs enterprise and returned when the last of the pigs went in 1974. They both love some form of livestock on the farm, so much so that they bought two in calf heifers for £200 each. At the time John farmed with his brother on 150 acres but some years later split the farm, John taking 75 acres on his own.

The cattle numbers slowly increased with the help of Jane's father who gave them two Friesian cows as a Christmas present. The Friesians are high milk-producing cows so were used for multi-suckling calves to rear for beef. Many farmers in the vegetable and potato growing areas of the Fens fattened cattle on by-products from those crops, the manure being returned to the soil for fertility. Little did the family realise that farmyard manure would be the mainstay of their farm in the future. Two of their sons, Robert and Christopher, followed in the family tradition of farming, while Richard went into Agricultural Engineering, being awarded the JCB Award at Writtle, and eventually joining the firm of JCB. Jane and her daughter-in-law Becci both manage the farm office, which in today's world is no mean task, and consequently know and understand the daily workings of the farm as well.

All cereals grown were eventually fed to the cattle along with silage making this the initial entry into an organic farming system, and cabbages were introduced on the Stickney farm as an organic crop. Growing organic crops did not prove difficult whatsoever, but marketing was another matter.

John was a member of Old Leake Growers, a growers' marketing co-operative formed to market produce from growers in the surrounding area, who at that time were not involved in organic produce. Undeterred, John and Jane decided to go it alone. They marketed their produce through various outlets involved in the 'box system' – a system where a variety of vegetables are placed in a large cardboard box and delivered to the doorsteps of houses in towns and cities – as well as selling at the farm gate. The Soil

Association had steered John successfully into producing organic beef and cabbages but he believed they were lacking on the marketing side of the business.

John made contact with Bill Allen, a cattle buyer for Organic Farmers and Growers, who established an outlet for their beef. The Soil Association later amalgamated with Organic Farmers and Growers and formed Organic Livestock Marketing Co-operative who market all John's cattle.

While the fertiliser and agrochemical bills are lower than if John and Jane were farming on the traditional method, their labour costs are much higher. The labour force consists of sons Robert and Christopher as well as three other local men and up to ten casual workers, both male and female, who are all from Eastern Europe. This is not unusual for casual labour on the farms in this area – without this nucleus of foreign labour the food growing industry would grind to a halt. No sprays for weed control are used on the farms as all weeds are controlled by mechanical means or by hand labour. Early cauliflowers, for example, are hand-weeded after the polythene covers are taken off, while all other vegetables are kept clean after planting and through their growing life by hand labour.

To maintain fertility in the soils for potato and vegetable crops, crew yard manure is used. Straw is baled and carted from the fields to the crew yards where cattle are housed in the winter, and used over the winter months as bedding for the cattle to lie on where it becomes mixed with the dung. The crew yards are cleaned out in the late spring when the manure is taken to the fields, after it has been piled into large heaps to decompose and become 'crew yard manure'. Land for potatoes is usually ploughed in the autumn and manure is spread on the fields prior to ploughing. This method of maintaining fertility is not cheap but neither are artificial fertilisers which are the basis of most farm crops and have more than doubled in price in the past few years.

Yields of most organic crops are on average half of a conventionally grown crop, except beans which are a leguminous crop and fix nitrogen via their roots in the soil for the following crop. The yield from beans is the same as those grown on a conventional farm because they require no artificial fertiliser to grow.

Ostlers Farm is totally sustainable and a very rare example of its kind, never having seen artificial fertilisers in its history, whereas the Wrangle land has been transformed from traditional farming to organic. The cattle breeding herd consists of 90 cows of pure-bred Lincoln Red and Aberdeen Angus, which are bought in from only organic herds. John and Jane consider these two native breeds excellent, but lacking on their

The cow in the centre of this 2009 photograph is a Lincoln Red crossed with a Limousin, noticeable by her good hind quarters as a result of the Edwards' breeding policy. *(RS)*

The Edwards feed milling and mixing plant in 2009 where home-grown organic cereals are prepared for their organic livestock enterprise. *(RS)*.

hind quarters compared to the continental breeds. For this reason they are put to a Limousin bull which does add some size to the rear quarters of their calves. Around 10 per cent of their heifer calves are retained to join the breeding herd and crossed again with a Limousin bull, so the next generation of calves are well fleshed on the hind quarters producing a good grade of meat.

Cattle from New Farm, Wrangle, go on to the nearby salt marshes during the summer months after which they return to the farm where they spend the winter months in yards. Feed consists of silage made on the farm, veg and potato waste and their own home mix of wheat, oats and spring beans, all milled and mixed on the farm. The arable cropping consists of spring and winter barley, oats, beans, grass silage mixture, spring, summer and winter cabbage, red and white cabbage and cauliflowers. Most of the vegetables and root crops are marketed through Lincolnshire Field Products who supply Tesco and the Co-op.

These comments were made by the family:

'In the present economic climate, organic foods have witnessed a decline in demand against traditionally produced products due to their higher price. If this is only temporary, we will survive this downturn; if, however, prices fall further and we could not make a living supplying organic food, we would reluctantly have to return to conventional farming.'

'Economics drive farming like any other business.'

The Countryside Stewardship Scheme was introduced to improve the environmental value of farmland throughout England. A farm entered under this scheme was required to maintain certain landscapes and features including dykes and ditches, wildlife corridors in arable areas using uncropped margins in arable fields. Management to benefit associated wild flowers and birds, and traditional buildings, old meadows and pastures for maintaining and increasing biodiversity. The scheme has one year to run (2010), when it will be replaced by the Higher Entry Level Scheme, or Organic ELS.

Peas

The Romans had a taste for the chick-pea but there is little evidence that peas were grown or consumed in the UK prior to the Norman Conquest. The early name for the pea we know of today was 'pease'. Due to its nutritious value and ability to be stored for several years, it was valued in times of dearth. It was an ideal source food for ships travelling long distances such as to the Americas, and in 1635 is found in a list of supplies for one year's journey on a colonist ship.

In Europe fresh peas became fashionable, and a delicacy to eat in the late seventeenth century. Arthur Young does not mention the crop in 1813 as being grown in Lincolnshire, but there is mention of them as an arable crop in Cambridgeshire by the Revd W. Gooch in 1811, referred to as 'pease'. Gooch refers to varieties as 'no sorts peculiar to the county, the white, grey and dun are common, being grown in open-field usually instead of barley or beans. They can only be kept clean when drilled in wide intervals, a plan the Cambridgeshire farmer has not yet adopted.' Harvest was by 'a pease make' which I assume was a rake, and then they were left in small heaps and turned as often as the weather may have made it necessary. This method of harvesting remained well into the twentieth century. Gooch does not mention storing or thrashing the crop.

They would have been grown mainly for human consumption as a source of protein with a long shelf life and when required, soaked in water until soft, then cooked. Wheeler in 1868 mentions blue peas being, 'grown following oats and succeeded by wheat, requiring clean condition of the land otherwise they become smothered with weeds.' Yields were 3–5 quarters to the acre, large crops in good seasons yielding 5–7 quarters per acre. This variation in crop yield illustrates its susceptibility to the weather and growing conditions, which has not changed to this day.

The surge in the fish and chip industry in the late nineteenth century created a demand for dried peas for the 'mushy pea' trade. This, coupled with the shortage of protein food during the First World War, was the birth of the pea as a major crop in the Fens which would last well into the twenty-first century. This precarious crop over the next decade would witness many changes for its demand, husbandry and agronomy in the Fens.

Harvest peas, which were peas that had matured in the fields, or stack, were the main source of this crop for the merchants and processors, from the late eighteenth century to the mid-twentieth century. Canning was initially with dried peas through to the 1950s after which fresh peas were canned followed by freezing in the late 1960s. Fresh peas in the pod became fashionable in the early part of the nineteenth century through to the post-Second World War years. Available transport by rail and road from the Fens helped stimulate this trade, as did the abundance of labour sourced locally and from the cities.

Green Peas being picked for the fresh market in Thorney Fen in the early twentieth century.

Kelly's *Directory of Lincolnshire* from 1922 lists six pea merchants in the county with mention of the varieties 'Blue and marrowfat peas', which would remain as the two main varieties of harvest peas to the present day. These merchants would have been trading in dried peas for human consumption and stock feed. Wherrys of Bourne at this period were major packers of dried peas for human consumption. Peas were already grown around the Bourne area and nowhere more so than the fertile Fen soils. They were an unpredictable crop to grow and suffered from blemishes and insect damage, requiring grading by hand. This job was labour-intensive – an ideal job for women folk who were available in the town and surrounding area. Harvest peas remain an important part of their business to this day.

By the early twentieth century canned foods were available in the UK and used extensively during the First World War. This period witnessed a new dawn in pea growing across the Fens, with fresh peas for canning alongside the traditional harvest varieties for processing, and fresh peas in the pod.

Canners were canning fruit, vegetables and fresh peas when in season, as well as dried peas for processing when lines were not processing fresh produce. One of the leaders in this field at that time in the Fens were the Smedley family. Before coming to the Fens they were fruit merchants in Evesham and in 1925 Samuel Wallace Smedley bought a fruit preservation factory from Crosse & Blackwell in Wisbech. Hearing how canning was developing in the USA he sent his son Wallace Venables, aged eighteen, there to learn how it was being done. He worked for several months there and returned complete with a new knowledge of the canning industry, along with a future wife whom he met on the journey.

In 1927, American canning equipment arrived as Smedleys imported the first mechanical pea viner to operate in the factory. After a few years, static viners were transferred from the factory to the farms where the peas were grown. In 1932 the National Canning Company PLC was formed who owned Smedleys Ltd, and would become a global enterprise.

Between 1937 and 1939 Smedleys were freezing fruits and vegetables, but in 1939 the government decreed all cold stores could be used solely for butter and meat. Freezing ceased until 1946, but canning continued.

In 1931 Smedleys had built their Spalding factory where they canned fresh peas along with their traditional lines of fruit and vegetables. Peas were brought into the factory

Fresh peas on the vine being brought into the Smedley factory at Wisbech for vining then canning in the 1920s. *(Lilian Ream collection)*

to feed the two static pea viners which stripped the peas from the vines. In conjunction with this operation, ten static viners were installed on farms at Long Sutton, Wingland, Postland, Gosberton and Bourne. Shelled peas from these viners were taken in boxes to the factory by lorries. With many materials hard to come by after the war, some firms used old ammunition boxes as pea containers. The output of the Spalding factory in 1950 was 5 million cans (200,000 cases) of peas processed annually, employing 200 regular staff and 50 students during the pea season. The canning factories were favoured by students during the summer holidays for seasonal work attracting UK and foreign students. Farrows of Peterborough had by tradition employed many Irish students, while Lockwoods employed many Greek students.

At about the same time, the Francis family of Boston built factories in Boston and King's Lynn, trading as Lincolnshire Canners known as 'Lin-Can', canning fruit and vegetables which included peas. Willer & Reilly were also a part their family business while Lockwoods of Long Sutton and Beaulahs of Boston all made up the growing numbers in this industry, canning many varieties of fruit and vegetables.

During the Second World War, large areas of grass were compulsorily put under the plough through government legislation and peas were an ideal crop to grow on land which had been down to grass for many years. With protein scarce at this time, harvest peas fitted this void either as processed, or sold as dried peas for soaking and cooking. Looking back to our farm records of back cropping during that period on newly broken up grassland, we were growing peas sometimes three years in the first four years of cropping. The crops were for fresh peas in the pod, pulled mostly by women and children for the fresh market, as well as harvest peas.

Peas from local farms being fed into the static viners for canning at J. Farrow & Co. of Peterborough in the 1960s. The lorry on the right has brought in the peas and the vined pea straw is being loaded on to the trailer on the left for cattle food.

Massey Harris 21 tanker combine harvesters combining peas which are transferred into trailers for J.W. E. Banks, Postland, Crowland, 1940s.

Prior to the Second World War, harvest peas were cut into windrows, turned, carted and put into stacks by hand labour, then thrashed as and when required for sale. Mechanical cutters did come in during the war and post-war period but the rest of the operation was carried out as it had previously been done. Due to inclement weather affecting the peas after cutting, some were put on to poles, and when required for thrashing or stacking they were moved by a tractor.

All canners and freezers of peas were using static viners on the farms until mid- to late 1960s. One company at the forefront of mobile viners was Food Machinery & Chemicals (FMC), now Pluger MC, who recorded their first trailed viner in 1946. When mobile viners came into use, the loss of pea haulm silage, a residue from the static viners,which was a high protein food for livestock, was lost. From then on the haulm was left in the field to be incorporated into the soil, which also had hidden benefits.

By the late 1960s canning was losing ground to the freezer, and the 1970s saw many canners and freezers taken over by the larger players in this field, such as RHM and Hillsdown Holdings who consolidated the industry. Many of the early freezing plants were set up by, Findus, Birds Eye, Ross Group and Eskimo Foods who primarily froze fish but added peas to their lines in the late 1960s. This was the period of EU-funded grants available from AMDEC for growers to form co-operatives for growing peas for freezing. Harvesting, marketing and vining groups were formed not only in the Fens but in other pea growing enclaves in the UK. Many groups were set up in and around 1967 when grant aid was available for mobile viners. FMC brought out their first trailed machine Type LV in the early 1960s, but it was their 963 trailed machine which made a impact on the harvesting of peas in 1966. The crop was cut and put into windrows by pea swathers to be picked up by the viners, which shelled the peas and returned the haulm to the field.

Almost all areas of the Fens and marsh by the 1970s were operating viner groups and the acreage of peas grew with many groups hiring land for one season to expand their production. This happened in other areas of the UK, mainly down the eastern side of

Three trailed pea viners being pulled by Caterpillar tractors on the front due to extreme wet conditions for the Postland Pea Group near Crowland in the 1980s. The peas will have been cut with swather prior to vining.

the country from the borders of Scotland to Essex in the south. A major change came in 1977 when FMC introduced their first self-propelled viners which eliminated the need to cut the crop prior to vining, which could be problematic during wet conditions. The new viners stripped the peas off the vines which then were fed into the machine itself.

The 1980s saw the vining pea acreage at its peak. Black clouds soon rolled in, with the processors having to compete with imports from the EU and its neighbours as well as other parts of the world. Fresh vegetables rather than frozen were more available from around the globe and becoming the consumer's choice along with the catering industry. The profitability of the crop from the post-war days onwards encouraged farmers to grow peas too close in their rotation, resulting in soils becoming infected with footrot diseases. This problem, unlike other crops where diseases resulted from close rotation could not be cured by agrochemicals – only a wider rotation was the cure.

The industry has slowly contracted over the past two decades and viners traversing the fields is becoming a rare sight, as is the entourage of mobile workshops, canteens and fuel bowsers. Residents would look out of their windows at fields of peas in the evening to wake up with the same field harvested and freshly frozen into '150-minute peas'. Many vining groups employed young men to carry out this work for the season, from Australia, New Zealand and South Africa who would then go harvesting cereals on other farms in the Fens, many remaining good friends.

Fen Peas Ltd, a growers' co-operative, was formed in 1968 under the AMDEC grant scheme by four growers, John Stephenson, Michael (Mick) Belton, Maurice Twells and Joseph (Joe) Pocklington, all farming in the Amber Hill and Holland Fen, area near Boston. It is the oldest farmer-controlled pea group in the country in their original format, with no takeovers or amalgamations with other groups. Stephen Francis, a member of the Francis family who were one of the pioneers of canning peas in the early 1930s, is still involved in the industry today with Fen Peas, as managing director. After leaving school in 1981 his first job, like many other boys and girls in the Fens, was working on farms. His first job was actually with Fen Peas collecting samples from the fields to be tested for their tenderness prior to harvesting with pea viners. The drilling period for vining peas is spread out between February and May, programmed on 'heat units' in the soil and different varieties. In theory, this should produce a continuous harvesting

FMC 679 self-propelled pea viners which replaced the trailed machines in 1977 eliminating the need for swathing prior to vining.

programme with peas for the freezing plants, but sometimes nature interferes. The end users require peas of differing tenderness, and so the samples collected by students like Stephen would determine when the viners would go into the sampled fields to harvest. The group chairman at that time was one of the founder members, Heckington farmer Bill Belton, who took Stephen under his wing for two years to act as field operations manager, becoming fully employed by the group in 1989.

At that time Fen Peas were growing about 600 acres of peas and running 5 LV trailed viners. In 1986 all the trailed viners were sold and a new FMC 879 harvester was purchased which operated 24 hours a day by two teams of operators and harvested 800 acres. In 1988 a FMC 679 was purchased and it harvested 1,500 acres. Two years later the group expanded to 2,500 acres and was running three FMC 879s. It was at this time that FMC (now PMC) were trialling their 979 six-wheeler, and Fen Peas were one of the first groups to buy one. Since that time the group has expanded its area to 4,000 acres with growers spread over a wide area in and out of the Fens. The harvesting operations are worked day and night by two shifts of operators, which are made up of viner drivers and cart operators. The carts take the podded peas from the viners and load them into lorries in or near the field. A fleet of lorries transport the harvested peas from field to factory, operating on tight schedules especially where strict profile of grades are required such as 150-minute peas – this is the time from the first pea being harvested into a lorry to the last pea frozen from that lorry load.

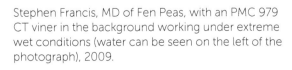

Stephen Francis, MD of Fen Peas, with an PMC 979 CT viner in the background working under extreme wet conditions (water can be seen on the left of the photograph), 2009.

The major changes Stephen has witnessed since his first days with Fen Peas has been in harvesting technology, which he believes may have reached its peak with this crop. The number of processors who freeze and can the crop has rationalised over the years. Factories freezing the group's peas today are at Bourne, Boston and King's Lynn with processed peas going to Premier Foods at Long Sutton.

From harvesting 30 acres a day in the early days, they now need to clear 100 acres per day using three PMC 979CTs which cost £365,000 each (CT meaning controlled thrashing, enabling the operator to make most adjustments from the cab). Vining peas are grown more for a specialist market today than they were in the past, with demands for various grades, and quality alongside organic crops for Waitrose, as well as peas for canning for Premier Foods of Long Sutton. The seasons change but the vining season has remained the same requiring lots of equipment, investment and expertise, together with good luck with the weather.

Soil-borne diseases are still a concern forcing the group to spread their wings further afield to find foot-rot free soils within a manageable distance from the freezing factories.

POTATOES

It is said by the fen people that fenland churches and abbeys were built on wool, and metaphorically this is correct; not the foundations obviously, but on the revenue generated from the product. This part of England was one of the prime producers of high-quality wool which was shipped from the fenland ports of Boston, Spalding Wisbech and King's Lynn to the Flemish weavers, especially between the eleventh and thirteenth centuries. The wool was of high quality produced from the forebears of our native Lincoln Longwool sheep. Individual crops such as bulbs, fruits and brassicas have also created wealth in localised areas of the Fens at different periods of our history.

However, having said all that, there can be no doubt that the potato, introduced into the Fens in the eighteenth century, has created more wealth than any other crop grown here. This crop was first cultivated in Scotland in about 1725 on a field scale commercially, and some thirty-three years later was mentioned as being grown in the Fens of Lincolnshire. A landlord's admission in Wildmore Fen in 1758 said 'that the potato crop was indeed of great value to the poor and of benefit in breaking up ground.'

Arthur Young in his 1813 *General View of Agriculture of Lincolnshire* mentions Brothertoft, Spalding, Butterwick, Freiston, Wrangle and Leak all growing potatoes, as well as Tattershall and Coningsby. These last two areas were on the Fen edge which, having gravel sub-soil, were less prone to waterlogging than the fen.

Reference is also made to potatoes grown on a large scale at Whittlesey, Chatteris, Soham and Upwell in 1811, and they were as profitable as any crop grown. Varieties mentioned are red nose kidney, rough red and ox noble, and yields of 80–170 sacks of three bushels per acre. Planting methods varied from seed placed in every third furrow, each about 4in apart, to rows 30in apart with seed at 24in apart. The seed potato was mostly cut into pieces the size of a hen's egg making sure each portion had an eye left in it to sprout. This method made economical use of a limited amount of seed purchase. Cultivation between the rows was done by hand-hoeing and horse-drawn cultivators. Harvesting was either hand-dug with three-tined forks or ploughed up and harrowed with horses. Women and children then followed to pick the crop into sacks. If the crop was being sold off the field they were sent off in bushel sacks. If they were to be stored over winter they were put into pits covered with straw and then covered with soil to protect them from frost. The pits were situated in a dry part of the farm which was free-draining. Liberal amounts of manure were advised for optimum yield and labourers around Whittlesey paid up to 50s per acre for the use of land to grow potatoes. Some farmers grew the crop and sold them by the acre for others to lift. In years of surplus the crop was fed to pigs, sheep and cattle with excellent results, while some farmers grew them only for stock feed. In years of flour shortage, potatoes were used as a substitute for making bread. They were an excellent crop in the rotation followed by wheat. However, mention is made of some farmers growing them for several years in succession. Mention is also made of crop disorders, distempers, curled roots and 'some sort of worm or insect eating the sets', causing a much-reduced yield. This was obviously the first signs of potato eelworm which was due to close rotation.

The newly drained Fens and marsh were the perfect soils to grow this crop – virgin land with no soil-borne diseases and of excellent structure. This area also has some of the lowest rainfalls in the UK, averaging around 22in per annum, which makes it less prone to potato blight.

One of the major drawbacks of growing potatoes is 'Phytophthora infestans', commonly known as blight, a fungal disease which spreads quickly through the crop in wet, humid conditions. It is common almost everywhere where the crop is grown, but some areas and districts are less prone to the disease than others. Likewise some varieties are more prone to blight than others and the weather is the determining factor in its spread between plants. The spores can also be carried over from old potato dumps, which act as a host, to the crops that follow.

After the potato's introduction into the UK, many impoverished families on the breadline were able to grow crops to provide sustenance during the winter months when food was scarce. Common land was readily available during the eighteenth and early nineteenth century in the Fens. Some areas of the UK were subjected to a 'Potato Tithe' but I have found no mention of them in the Fens. Much of the newly drained fen in the seventeenth century was deemed to be free from tithes, but not all.

1845–51 will go down in history as 'the blight era'. It was a major disaster for the potato crop with severe blight all over Europe and the great potato famine in Ireland. At that time little was known of the disease and certainly no preventive measures were in place to control it for another 40 years. This tiny fungus was responsible for the death of around a million Irish and forced the migration of millions of others around the globe, a reduction which saw the country's population fall from 8 million to 5 million in a short period. It could be said the humble potato played a major part in changing the course of history, certainly in the USA, where 43 million Americans claim to be of Irish descent.

S. Graham Brades-Birkes in his 1951 book *Modern Farming* extols the influence the potato has made in British history, 'It is probably true to say that long wars would be impossible without them and readers who have lived through two world upheavals must be aware of our reliance upon the crop.'

Access to the markets before the advent of the railways restricted the crop's potential in the Fens. The main railway line from Peterborough to London came in 1863 and many branch lines soon followed. By the 1870s almost every small town in the Fens had rail links to the major national routes and hence to the

Drawing out potato rows by horse, planting by hand and covering with horse and plough at Pickard Farm, Horbling.

Potato picking with women and children filling their baskets while men empty them into the cart, early twentieth century.

entire UK. White's 1882 *Directory of Lincolnshire* says that 'Potatoes are also becoming a very general crop, but the produce is somewhat uncertain in yield and quality.' Also, in some areas 'the quantity of potatoes now grown is astonishing!' Potatoes were an important crop now, with the railway links able to transport them anywhere in the UK. According to the *Spalding Free Press* of 26 January 1909, up to 1,000 tons of potatoes per week were being handled at Holbeach station. Most were sent to the London market in 8-stone Hessian sacks, but large shipments were destined for North America in 12-stone sacks.

Hovenden Hall near Holbeach Hurn, a Cheshire Home foundation, was built by A.H. Worth in 1910 with money made from his potato crop. The family have remained at the forefront of potato growing to this day. Worths developed a horse-drawn sprayer, manufactured by Grattons of Boston to apply Bordeaux Mixture to potatoes for the control of blight. Also around this period Arthur Saul of Leverton Hall, Boston, made his fortune growing potatoes, one year growing 999 acres of the crop.

Listed in *Kelly's Directory* of 1922, there were 55 potato merchants in the Lincolnshire Fens, along with potato chitting box manufactures, potato chitting house manufacturers, and potato riddle makers. Almost all the potato growers listed in this *Kelly's* were in the Lincolnshire Fens. Those listed must have been substantial growers, since the number of growers listed is less than the number of merchants listed. One would assume from this that there were very many small growers (not listed) who accounted for a large acreage of potatoes in their own right.

Potato merchants are not listed separately in *Kelly's Directory of Lincolnshire, 1933*. By then there is an entry for potato spray manufacturers (fungicides) which shows by then agrochemicals were already being used on the potato crop. At the turn of the century the stage was set for the Fens to turn into a potato Mecca. As the wholesale markets around the country, influenced by the general public, knew they could buy potatoes from the Fens, the scramble was on for the perfect potato. Different markets around the country developed a taste for different varieties. The flavour of a potato does vary from one variety to another and buyers became aware of this and selective. The soil a potato is grown in can also influence flavour and varieties were categorised as 'reds' and 'whites'. With soils in the Fens varying from several grades of silts, tofts, skirt and various peats, this produced a potato for everyone's taste. Many wholesale markets only wanted certain varieties, grown on certain soils as it became their trademark. Even in London, Covent Garden preferred potatoes grown on the silt lands whereas Brentford wanted King Edwards (a white potato with flecks of red skin) from the black peat land, as did the Birmingham market. The northern part of England had a passion for fish and chips and required a potato which would cook well and not break down when fried. Local potato merchants prospered in every town and village across the Fens accompanied by merchants from the cities establishing their buyers and warehouses here. Virtually all potatoes were transported by rail until after the Second World War.

The Second World War was the impetus to grow more potatoes which were for many families a major part of their staple diet. Large areas of the fen under old pasture was turned over to arable farming. Initially in 1939 a grant was paid to farmers to plough old pasture up for arable cropping, together with a subsidy for lime and basic slag (potash), to improve the soils. This was voluntary but it soon became compulsory under the Plough Up Campaign by the government. The campaign was not welcomed by all farmers in the Fens; my own family rented a heavy grassland farm which the owner would not plough up – consequently they did. Between 1939 and 1945 the potato acreage in the UK increased 116 per cent.

Most of the grassland brought under the plough was the lower, poorly drained heavy fenland. Much of the black peats until now were poorly drained but after extensive draining they also were turned over to arable. Drainage had become a government priority with many of the currently existing pumps built in the early war years. Draglines became commonplace for cutting new dykes and the American buckeye pipe layers had come into this country. Between 1940 and 1944, 4 million acres were drained, a great part of which was fen and marsh land. Much of the arable land had reverted back to grass in the late 1920s and early 1930s when prices of arable crops tumbled.

With the advent of the Second World War, land was turned back to arable and by 1944 the grassland acreage was at its lowest in English history. Tractors, mainly from the USA, became commonplace on fen farms, but the horse was still supreme. He was essential during that period of war, bred on the farm, fed from the farm, as well as adding to its fertility.

Potatoes were a labour-intensive crop and, with so with men away at war, the government created the Women's Land Army who filled a vital labour gap before the Italian and German POWs arrived. As mentioned elsewhere, as war progressed many other POWs came to the UK and worked on the land.

During the war agronomy made rapid advances but the plant breeding of potatoes did not. Many old varieties of potatoes were still being grown, such as King Edwards, Doon Stars and Majestics. Potato prices during the war were at a controlled price; there was a flat payment for all potatoes and they could only be sold to a local merchant for a fixed amount. The merchant was also controlled on his margins so it was not a free market. One example of this is Wharns Tinkler, a small-holder who farmed 24½ acres on black fen rented from the Huntingdon County Council. In 1941, Tinkler sold his White Doon Stars at £5 5s per ton in February. The Majestic potato was a very good potato for storing. That year he sold his Majestic potatoes on the 'July Reserve' where he received an advance payment in March for the potatoes to go off the farm in the following July. There were 9 tons of Majestics on account in March at £7 a ton, equalling £35. The balance when sold in July was at £3 2s 3d a ton, equalling £28. Total income from those potatoes was £73, the equivalent of £8 2s 3d a ton. Other prices during the war years were (King Edwards being red potatoes):

1939 Whites £3 12s 6d	King Edwards £4
1940 Whites £5 5s	King Edwards £7 5s
1941 Whites (Doon Star) £5 5s	King Edwards £6 6s
1942 Whites (Doon Star) £5 15s	King Edwards £6 10d

A gardening class at Fen School, Puddock Drove, from about 1920. Wharns Tinkler, Fenland small-holder, is third from the left, front row. *(R. Tinkler)*

Wharns Tinkler, fenland small-holder with his first horse, Trimmer, in the 1940s, who he worked from 1934 to 1943. *(R. Tinkler)*

To encourage farmers to hold on to their main crop potatoes until the first earlies were lifted, usually around July, a subsidy was paid to growers called the 'July Reserve'. Part payment was made to the farmer when he would normally have sold his potatoes, the balance being paid when they actually went off the farm. The Ministry of Food subsidy for 8 acres of potatoes paid a total of £80.

During the Second World War, Tinkler's potato crop accounted for a third of his total acreage. He was therefore growing potatoes every third year which was far too close in the rotation, but not unusual during wartime in the Fens. His sugar beet acreage was almost one in three years, also too close.

Tinkler farmed the same acreage (24.5 acres) from 1935 to 1957 when he retired. During that period his potatoes produced an average of 54 per cent of his total income, illustrating the reliance he put on his potato crop. Tinkler was not on his own; this was common in most districts across the Fens especially on the best potato growing land, but there would be a price to pay for it. By the mid-1950s much of the land in the Fens where a very close rotation of potatoes had been grown were suffering from potato cyst nematode, commonly known as 'potato sickness', later known as potato eelworm. Michael Thompson remembers his father's three-course crop rotation on their Holbeach Marsh land around 1942⅔ as being potatoes, sugar beet and wheat. By the late 1940s he was only growing potatoes on 5 per cent of the farm instead of 33 per cent, due to eelworm.

One of the first mentions of eelworm disease I have found was in the *Wisbech & District Fruit Growers Year Book* in 1945. It had been found in land growing narcissus as well as onions, an alternate host, where close rotations on both crops had been practiced. It was widespread in areas around Wisbech, Upwell, Parsons Drove, Whittlesey, Littleport and near Ely. Most of this land was, or had been some, of the best land for growing these crops, yet there was no mention of potato eelworm which is surprising.

The drive to produce more food, and income after so many lean years changed farming from rotational husbandry to an almost mono-culture system. The nation had to be fed at any cost whatever the detriment to the soil. However, not all farmers felt this way. To many traditionalists it was draining the last reserves from their precious soil, seeing the old pastures broken up and farmed extensively. The old adage of 'live for today but farm as if you are going to farm forever' was not heeded. It was the time in our history when no one knew the future, a precarious period to live

Tinkler's sales for 1935 from his 24-acre CC holding in Warboys Fen. Note that potatoes were over half his total income.

in, but the soil had to be farmed to its limits as there was no other option. It was not only the potato which has been subjected to intensive growing; other root crops as well as peas had infested the Fens with soil-borne diseases.

Trials were carried out in 1949 on several silt and black soils where eelworm was a problem, injecting a DD-mixture into the soil before planting potatoes. It proved successful on the silt soils but not so on black soils high in organic matter. Chemicals such as Vond, Metam, Vidate, Telone and Temix were all used to treat soil for the control of potato eelworm. Today Nemathorin is used, but with present legislation looming to curtail chemicals to control soil-borne diseases, this may soon be taken off the market and alternatives found. One crop being tried is Braco, a resistant mustard sown in August and September and ploughed in before Christmas, which breaks down the eelworm life cycle before potatoes are planted in the spring.

It was not only intensive cropping of potatoes during both world wars that increased eelworm; one farmer I know who grew potatoes on a wide rotation also suffered with the disease. In the post-war years up until the 1970s most fen farms had livestock in yards in the winter months. This farmer, like many others, bought stock feed potatoes from potato growers on the silt soils, where potatoes had been grown intensively to feed to his cattle. He believes they were his source of infection.

The demand for more food from British agriculture would remain into the 1960s and to satisfy this demand only modern technology could help to revive our soils. Agrochemicals and fertilisers had made untold advances during the post-war years in all sectors of agriculture and the likes of guano, basic slag and ammonium nitrate have been replaced by blended compounds to suit soil deficiencies and the requirements of individual crops. Had it not been for the advance in agrochemicals to control this disease, large areas of the Fens would have ceased growing potatoes on most of the best potato growing soils in the Fens. The seriousness of this legacy still hangs over this area with the disease still widespread. Growers have, however, been concerned over this issue and are extending their rotation to reduce the problem.

Potato blight was still the major concern, especially during wet periods in the summer months. Chafers were leaders in this field with their crop dusters for applying copper oxychloride (Dithane 945). Copper was superseded by a zinc- followed by a tin-based material. A new word in the agrochemical industry appeared which would change the whole industry, systemic fungicides and insecticides. These were sprays, or granules applied to a crop which took up the chemical in to its tissue system, to combat insects and diseases.

Outbreaks of blight could also be treated with new fungicides from the air. Companies such as S. & S.G. Neal commenced spraying with helicopters, as Fisons Pest Control, while names in the fixed-wing business included Lincs Aerial Spraying and Hardy & Collins, both of Boston. Both methods of applying insecticides and fungicides were perfected to cover thousands of acres when the need arose, especially in blight years.

The late 1960s witnessed a shift of labour from the farms into industry in the towns and cities around the Fens. The Irish labour force coming to the farms declined and women who had traditionally picked potatoes took jobs in the pack houses and processing plants. Mechanisation on the farms was progressing in the 1960s but this change in the availability of labour advanced it especially with root crops. Potato harvesters made a limited appearance in the late 1950s but by the end of the 1960s most crops were lifted by this method. Johnsons of March and Whitsed of Peterborough were pioneers in this field, Johnsons also developing potato planters and other machinery for this crop. The sight of Irish labourers toiling in the potato fields had gone and their hut accommodation became a home for the jackdaws. Horses munching their morning feed at 6 a.m. ready for the day's potato picking and returning at night fall were no longer to be heard. The footprints of both were never seen again on fenland soils; it was the beginning of a new era for the humble potato.

Potatoes in paper sacks being loaded onto a Ford flatbed lorry after being riddled in the 1960s.

Up until this period almost all farms in the Fens grew potatoes, whatever the soil type. Machines could not harvest potatoes on heavy soils because of too many clods, so hand labour was the only method of picking. Potato growing on heavy soils in the Fens virtually ended when handpickers became unavailable. Our family ceased growing potatoes on heavy land in the late 1960s.

New varieties of potatoes were introduced for the packing, chipping and processing industry, doubling the yields. Planters, lifters, graders and packers took the potato through every stage of its life while agrochemicals, applied with both land and aerial sprayers, cared for its health. Technology had fulfilled its purpose.

Entry in to the Common Market in 1972 brought the era of farm co-operatives. Farmers were encouraged and financed to form these co-ops to pre-pack and market potatoes. Few of these co-ops remain today as farmer grower organisations; those who do now belong to larger marketing groups through management buyouts.

The frozen chip originating from Canada in the early twentieth century became synonymous with Birds Eye. The 1950s was the decade of frozen foods, brought about by advertising when people wanted convenience foods to prepare when watching TV. 1968 transformed the Fens as far as frozen chips were concerned when McCain built a factory at Whittlesey, the largest chip factory in the UK.

The late 1960s was the time flavoured crisps became the fashion and a surge in the market increased the demand for Record potatoes, a suitable variety for crisping. Pentland Crown was a new variety which spread quickly because of its high yield potential. These high yields depressed prices on the open market, but undeterred, farmers kept planting. Overseeing this transformation was the Potato Marketing Board (PMB) who controlled the acreages with quotas and in years of plenty bought up any surpluses. These surpluses were used for stock feed or just left to rot and spread on the land.

The supermarkets by the 1980s had changed the entire marketing strategy of the potato industry, and now accounted for the major slice of the retail market. Crisps of every flavour along with many new potato-based snacks took their place on the supermarkets' shelves and in pubs and hotels. Potato products were now sold in shops, filling stations – in fact, almost everywhere the public ventured. Most of the potatoes for the retail sector were being washed and pre-packed. An advert in 1969

Lifting potatoes destined for McCain's factory on Roger and Debbie Hunt-Payne's farm, North Fen, in 2009. *(RS)*

in the *Isle of Ely Farmers Journal*, was for 800–1,000 tons of potatoes per week for pre-packing peeling, processing and marketing by David Johnson (farms) Ltd, White Fen, Benwick, March.

Supermarkets who originally bought from the wholesale markets now bought direct from farmers' co-ops or the few remaining merchants who pre-packed potatoes. This era saw the demise of the wholesale markets along with the country merchants.

Transport of Potatoes

The railways had been the catalyst for the growth of the potato in the Fens for many years. Seed came from Scotland to the Fens where they were grown into potatoes and sent to all parts of the UK. During the late 1940s and early 1950s lorries – many of which were ex-Second World War Army lorries – began to transport potatoes to the cities. My family were potato merchants and growers at that time and their first lorry was an ex-army lorry. This would be the demise of the rail links in the Fens and by the late 1960s all branch lines had ceased. They had done their job when they were needed but this was the birth of road haulage.

Potatoes after being lifted and put into Dickie Pies for storage on A.H. Worth's farm at Holbech Hurn, 1960s.

The Labour government had nationalised road transport under the banner of 'British Road Services' in 1948 which encouraged many merchants to own their own lorries for transporting their own potatoes. BRS changed its name in 1969 to the National Freight Company, before eventually being denationalised in 1982.

Potatoes were handled and transported in 1cwt hessian bags up until the 1960s after which 4-stone paper bags became the norm. Eventually, with the packing industry growing, bulk transport became the way to move large quantities of potatoes. Storage of this crop was in mainly potato graves, or pies, until the 1950s. These were triangular heaps covered with a thick layer of straw then covered with soil where they would remain until required for sale. The 1950s also saw the introduction of the 'Dickie Pie' to the Fens. This was a rectangular clamp of baled straw with a ventilation duct laid down the centre on which the potatoes were placed with an elevator up to a height of around 8ft, then it was all covered with polythene and straw. Large tonnages could be put in these pies, some many yards in length. Specialist bulk stores were being built in the 1950s through to the 1970s. As quality became a keyword, the crop changed to storage in boxes and temperature-controlled stores to maintain quality throughout the storage period. Today potatoes can be stored in controlled stores until the next harvest.

PREPARING POTATOES FOR MARKET FROM STORE

In the days of potato graves, rockers were used to grade and clean the crop, commonly known as 'riddling potatoes'.

The first machines were hand-powered but later by small petrol engines operated by a gang of men or women who fed the machine, picked off the waste and weighed them into bags. Hessian 1cwt bags were used until the advent of potato bags weighing ½cwt. This all died out when bulk storage took over and grading was done in store. Today only a small amount of the crop is graded on-farm, most going to central grading packing and processing stations.

The number of growers has declined steadily over time. We are seeing either very small growers growing for the specialists or niche markets or the very large growers growing several hundreds, even thousands of acres. The cost of growing and equipping a potato growing enterprise is capital-intensive. The marketing is far more important now than in the past with the dominance of so few outlets for the crop. This has concentrated the entire industry into a fewer hands. Smaller farmers who once grew potatoes now let their land to the larger growers, so the total acreage grown in the Fens has not reduced.

A gang of men riddling potatoes with hand riddles and bagging in the early twentieth century. A hand riddle is on the stand in the centre of the photograph.

Machine collecting trickle irrigation pipe before lifting potatoes on Roger Hunt-Payne's farm near Whittlesey in 2009. *(RS)*

Tractor cultivating prior to potato planting and applying agrochemicals in 2009 on Bill Burton's farm.

The potato is still as important as it ever was to the Fens. Agrochemicals, growing, lifting and storage, have each played their part in maintaining its dominance as a farm crop. In the Fens 62,000 acres are grown annually, which accounts for 24 per cent of the entire production of the English crop. Around this crop, canning, freezing, packing and processing by some of the largest businesses in the UK survive.

The supermarkets and processors have changed the whole attitude to potatoes. Their unrelenting drive for quality for their customers has without doubt changed the entire industry. Traceability of the crop and awareness of the environment they are grown in have been at the forefront of their demands. These issues have been retailer-led, not producer-led. Like many other changes in history it has come at a price, that being the demise of the small man, both grower and merchant, which would have come with the shifting sands of time anyway.

Markets in the past were influenced by the weather and yield. Today there is no 'weather market', when prices fluctuated daily. Surpluses are rare as production is geared to demand and potatoes can be sourced from Europe in a short space of time if required. The supermarkets and processors have stabilised the market and as a result growers have streamlined their production, grading and distribution – some would say to the detriment of the grower, though this was not evident in my research.

Our soils have at times been farmed to extreme limits to produce potatoes, but thanks to man's ingenuity it has remained a premier crop in the Fens. With the loss of suitable potato land through development, urbanisation and conservation, if we are to fulfil the demands of the public for this crop, our soils will be put under pressure gain. Next time our soils will have to look once again to the scientist for an antidote which may have to be the genetically modified potato, if future generations wish to eat potatoes. To grow this crop to the standards which are required, irrigation is playing a major role and it may reduce the acreages grown on land where it is not available.

SUGAR BEET

The sugar industry from the seventeenth to the twentieth century was dominated by the sugar cane producers of the West Indies. The import of cane sugar into Europe during the Napoleonic Wars (early nineteenth century) was severely disrupted and as a result sugar beet began to be grown there. By the early twentieth century the industry was well established in mainland Europe. Britain at that time still had many cane refineries supplied by its colonies and was also importing beet sugar from Europe. As a result we were reluctant to venture in this industry. It reflects the attitude our governments had at that time towards British farming, when it was in the worst recession it had ever been in. There was a movement in 1909 by the National Beet Association to establish a 'home industry' supported by the Lincolnshire Beet Sugar Company who were intent on building a factory at Sleaford. Unfortunately, insufficient capital was found and it did not materialise. Dutch interests built the first factory at Cantley in Norfolk in 1912. The First World War exposed the vulnerability of imported sugar and so more factories were built after the war. Eighteen factories were up and running by 1928 under the management of five groups, due mainly to the Sugar Industry (Subsidy) Act of 1925.

Under the Anglo-Dutch banner our first factories were built in or near the Fens at Ely, King's Lynn and Spalding under the Anglo-Scottish Group, and Peterborough under the Bury Group. Wissington Group built their factory on the Norfolk fen edge.

Beet processing factories required important sites to operate. They required an abundant source of water for processing and conveying the beet around the factory. Rail and road links were also essential to bring beet in to the factory, and sugar, molasses and beet pulp out of it. The Fens had these assets and became major players in the sugar beet industry for processing and growing beet.

Almost all areas and soil across the Fens were ideal for the beet crop. Silts and fen edge soils were excellent while the heavier land grew less acreage of beet. The black fen soils grew good crops but had problems with 'fen blows' when equinox winds blew in the springtime, sometimes destroying the young seedlings. Straw was sometimes planted between the seedlings to act as wind breaks.

Roger Leonard, farms manager for Shropshire Farms, showing where straw has been planted in rows to protect seedlings from a fen blow in the spring of 2009. *(RS)*

Poplar trees seen in 2010, planted as wind breaks on peat fen soils.

Another method to combat wind erosion was to drill a thin crop of spring barley early in the season on the field due to be drilled with beet. The beet would be drilled when the barley was a few inches high and when the beet seedlings were well established the barley would be desiccated with a chemical which did not harm the beet plants. Hopefully by that time the beet plants were strong enough to survive any more winds after equinox time. The control of weeds was also a problem using residual sprays but was overcome when the correct contact herbicides came onto the market. Lime deficiency on organic soils was a problem. A valuable by-product from the beet factories was sludge lime which became a very valuable source of lime for lime-deficient soils.

Until the late 1960s, most fen farms fattened store cattle and sugar beet pulp became a valuable source of feed. All in all, the sugar beet crop was a boon to fenland farmers, and still is.

The UK factories were incorporated by statute into the British Sugar Corporation (BSC) in 1936 and the national beet area was 140,000ha or 345,000 acres. It is difficult to be specific as factory boundaries have changed over the years, as have throughput and grower numbers. A conservative estimate would be a figure of 75,000 acres grown in the entire Fens at that time. Over-production took place around the globe with falls

Sugar beet being topped and loaded into horse carts after lifted by steam traction equipment, on Mr Smith's farm.

in world prices and dumping of surpluses making our home-grown industry unviable without subsidies. The government issued a white paper in 1935 guaranteeing support for the industry and proposed the creation of the British Sugar Corporation in 1936. From that time, growing, manufacturing, refining, marketing and research were under government control. They also would subsidise the industry. The British sugar industry

Lifting sugar beet with steam traction equipment on Mr Smith's farm, Monks House, Spalding, in the early twentieth century.

Sugar beet being loaded into a cleaner by hand which lifts it into a lorry at John Saul Farms, Leverton, Boston, 1940s.

served a valuable service during the Second World War supplying the needs of this country for sugar and animal feed from its by-products. Sugar was rationed during the war and sweets remained rationed until February 1953. The Sugar Act of 1956 replaced the 1936 act which ended state trading in sugar and set up the Sugar Board. Its function was to buy Commonwealth sugar under the Commonwealth Sugar Agreement at a subsidised price, and sell it commercially. A subsidy was maintained which was necessary to fund the guaranteed price paid to growers for sugar beet, without which the industry would have ceased to exist. Lessons had been learned from the government's attitude towards British agriculture after the First World War when it slid into recession in the late 1920s and early 1930s. The industry has survived to this day after reductions of sugar quotas by the EU and a takeover by Berisford and Associated British Foods. It was a vital part of our agricultural industry and, with sugar prices rising globally, it was proving itself an asset to the nation.

In 1980 the national crop area was 210,000ha or 520,000 acres. As yields have gone up from about 35t/ha (14t/ac) in 1980 to 70t/ha (28t/ac), the area of beet required has steadily fallen, while the amount of sugar produced has remained broadly similar.

Factory closures have been inevitable during those turbulent years, beginning with Ely in 1981, Peterborough in 1988, Spalding in 1991 and King's Lynn in 1994. In 1980 the original five factories sliced 2.6 million tonnes of beet (about one-third of the company total). The only remaining fenland factory in 2009 is Wissington, whose campaign alone sliced over 3 million tonnes of beet. The area of crop grown today processed by Wissington's 1,350 growers, which includes some highland parts of Norfolk and Lincolnshire, is now 40,000ha or 100,000 acres. The National Farmers' Union celebrating its centenary published a leaflet, *Why Farming Matters in the Fens* in which it estimates the acreage of sugar beet grown across the entire Fens in 2008 was 53,000 acres which accounts for 17 per cent of the English crop. There have been many dramatic changes since the first fenland factory took beet from its farmers in 1928. Growing, harvesting, agronomy and transport from field to factory have changed out of recognition.

Genetic monogerm pelleted seed, introduced in the mid-1960s, reduced hand labour requirements for singling seedlings. The crop which once required an abundance of labour during its season is now mechanised throughout its life. Gangs of men and women were seen working alongside each other, hoeing and singling, chattering as they moved like skeins of geese up and down rows of beet.

Transport has moved on too, from barge, rail and horse and cart to 40-tonne lorry loads. Harvesters now lift vast areas per day working day and night on low ground

Wissington sugar processing factory on the edge of the Fens.

Dawson Bros of Bicker lifting sugar beet with a Vervaet 9 row harvester, which has a tank capacity of 25 tonnes, 2009. *(RS)*

pressure tyres with minimum ground compaction. It is still a crop in the Fens where small growers survive with 5 acres alongside large agribusiness growing hundreds of acres. The crop is once again in demand with more acres being grown here on varying types of soils. In many areas of the UK, sugar beet tops left on the fields after lifting are grazed by sheep, but in the Fens most are returned into the soil. The crop in the southern Fens attracts winter visitors of Bewick and Hooper swans by the thousands to graze the beet tops, while some areas nearer the marshes attract the Brent and Pinkfoot geese in vast numbers. The factories had their own sports clubs and social clubs frequented by their seasonal and regular workers, all of which added to the social life in the Fens. My memories of the Spalding factory, which closed in 1991, are that it was my weather forecast – if I could smell the beet factory during the campaign season, it would be cold with the wind blowing from due north. The smell of molasses, however, had a warming odour to it.

CEREALS

Evidence shows that parts of the Fens have been growing cereals from the Iron Age, Roman and Norman period. A large part of the Fens area before the seventeenth century was uncultivated land supporting only pastoral agriculture, tillage being carried out on the higher drier areas of silt and gravel islands. These islands are now the centres of population, being either towns, villages or hamlets. Even after initial drainage many of the soils were highly organic and too rich for cereal growing, requiring crops such as cole seed which was sown after paring and burning. Wireworm and mildew were also a problem growing cereals on fen soils.

The cereal varieties grown at that time, if grown on very fertile soils, would have tended to grow an abundance of straw, lodging on the ground and making harvesting difficult and prone to disease. It was a common practice if wheat crops were too lush in

A gang of women working on Mr Walton's farm at Fleet Coy, Gedney Hill, in the early twentieth century.

Binding wheat with three horse-drawn binders on Mr Edwards' farm, Wrangle.

Three binders cutting wheat on John Richardson's farm at Twenty near Bourne. Seen left to right are A. Asher, S. Webb, C. Baker, H. Fletcher, R. Parrish and R. Beddoes, 1960s.

the spring to graze them with sheep. This reduced the plants and gave it more standing power; it was also a valuable feed in the spring months when grass was in short supply, a practice our family still carried out in the 1970s.

On the newly drained organic soils frost tended to lift the soil in the winter months and damage the roots of the wheat crop. The Revd Mr Gooch in his 1811 book *General View of Agriculture of the County of Cambridgeshire* says, 'In the fen if much frost during the winter months the wheat is trampled in spring by men (three or four) abreast, price 4

One of the first tractors pulling a binder in Thorney Fen on Mr Burton's farm in the 1930s.

Thrashing machine driven by steam traction engine on Mr Edwards' farm, Wrangle.

shillings per acre.' He also states that harvest is, 'reaped, and generally by "acre men", mostly strangers, Irishmen who come over in large companies, and do that work only.'

Mr Edes of Wisbech at that time said that wheat in the Fens is apt to be very 'strawy and full of blacks' (this could be Ergot, or Takeall, both fungal diseases in wheat ears). Wireworm was a common problem which occurs mainly on newly broken up grassland, as was mildew, normally prevalent in wet, moist growing conditions. Gleaning of cereal ears left on the ground after harvesting by the poor in the parish was a common practice at that time, but many farmers considered it an evil. The crop would have been reaped, tied into sheaves and left in the fields where it remained for days or weeks depending on the weather and availability of labour to clear it. Gleaning was not permitted while the stooks were in the fields, but people did sometimes glean between the stooks of corn. Local acts in some areas gave magistrates the power to fine or imprison if the poor gleaned the fields before the crop was cleared and a penalty of 10s or 1 month's imprisonment was charged for this offence. Farmers were also liable to a fine of 5s an acre if they turned any stock onto fields before an appointed number of days after the crop had been cleared. This was a period of grace to enable the gleaners to glean the field or corn ears.

Wheat was the predominant cereal crop grown in the Fens with a limited amount of barley and, to a smaller degree, oats grown. All operations such as drilling, hoeing, harvesting and thrashing were carried out by hand labour into the mid-nineteenth century.

CEREAL SALES AND MARKETING

Marketing agricultural produce as we know it today was conceived in this century. Until this period a farmer would sell his corn and pulses to a local miller or salesman who acted as his agent. After thrashing, corn was put in sacks, usually 240lbs. He then sold it by two methods, 'pitched market' or 'sample market'. Pitched was where he took a load of corn to the market to be sold. The buyer was able to see what he was buying and this often encouraged him to offer a better price at that time. The seller, however, had the risk of not selling, carting it home, and returning the hired sacks. Many farmers were a long distance from the markets and not accessible on hard roads which could be arduous in wet weather. A 'sample' market was where the farmer took a sample of his corn to a market, where the buyer bought on the good faith that it was a fair sample of what he would receive later. Mostly the seller would offer less, but where the farmer's integrity was guaranteed he may offer more. The buyer kept half the sample, the farmer keeping the other, as a kind of insurance. The sample method would carry on through the nineteenth and twentieth centuries and became the trading principle we have to this day.

Samples were tested by the merchant or miller by his nose, for mustiness, and his teeth for its moisture content, or hardness, as it became known. Corn was harvested at all moisture contents, determined by the weather at harvest. It would be stooked in the fields to ripen and carted to the stack yards, ricks or barns where it would dry. Before the advent of the mechanical thrashing machines during the mid-nineteenth century, corn was thrashed by hand, a labour-intensive task, or a mechanical thresher driven by horses. Sometimes the top of the corn stacks would be wetter when thrashed than the middle or bottom. When this occurred those sacks of corn which held the wet corn were tied at the top with a wisp of straw to allow moisture to escape. The miller knew then which sacks to mix with the dry ones.

The advent of the steam engine revolutionised agriculture and milling. Steam engines began to power thrashing machines and mills, as well as providing transport by rail and road. Steam cultivators for ploughing and cultivating enabled farmers to bring vast areas under arable rotation previously used for grazing. The early to mid-nineteenth century would become the golden era of the corn industry in the UK, and ironically the latter part of the same century would prove to be the worst in agricultural history.

Modern farming practices, improved drainage and more land in agricultural production have increased the farmer's crop yields. Steam power was available for flour and provender mills, as well as water pumps and thrashing machines which set the scene for a new era in agriculture. Labour had left many of the farms to earn higher wages in the cities and towns, but with new machines on the farms to bind, thresh and handle corn crops, it was of little concern. The horse remained the main source of power on the farms until the early twentieth century when they were replaced by tractors, wheeled and tracked. The Fens were renowned for the breeding of Shire horses with some of the largest breeders in the UK. With so many in the area it is understandable why they were still being used so widely into the 1950s and even into the 1960s.

The enormous growth in grain and pulses had to be handled and marketed. The main grain trading house of the UK was in London in Mark Lane, with Liverpool and Edinburgh also having major corn exchanges. Here they would handle grain from around the globe as well as home-produced grains; it was the hub of grain trading that would have an effect on every farmer who grew cereals and pulses. The first building at Mark Lane was erected in 1747, replaced with the second in 1828, trading on Monday, Wednesday and Friday. Bristol Corn Exchange is the only surviving eighteenth-century exchange and is a remarkable building of outstanding architecture. To feed the likes of Mark Lane and other grain trading exchanges in the main cities, the demand for regional exchanges grew, with almost every market town in the country witnessing one being

Three Suffolk Punch horses pulling wagons belonging to Mr Hoyles of Tydd St Mary, Wisbech. The stack yard is behind with stacks thatched for the winter. *(Lilian Ream collection)*

built. They were situated in the centres of cities and towns and built on a grand scale befitting a theatre or opera house, which ironically several of them would later become. They would not only serve the agricultural community but the general public as well.

It was not only grains that were marketed in these buildings. Every commodity from insurance to fertilisers and from produce to machinery were traded there, as well as gossip and scandal. They were used for functions of every description both social and business other than market days. In the fen towns they were in the most prominent position in the town, and those that are still standing are a reminder to future generations of the birth of marketing agricultural produce, primarily corn.

During the infancy of the corn exchanges (1840–50) the price of wheat remained relatively stable at around 52s a quarter. Railways were the great windfall to British agriculture, opening up the countryside to the ports. The same was also happening across the United States enabling the prairie farmers of the Mid-West and Canada easy access to their east coast ports for shipping to the UK. Most countries importing wheat in Europe decided to levy tariffs on imports.

The interior of Peterborough Corn Exchange, built in 1848 and demolished in the 1960s.

Top left: Bourne Corn Exchange, built in 1870.

Top right: King's Lynn Corn Exchange, built in 1854.

Centre left: Long Sutton Corn Exchange, built in 1856.

Centre right: Wisbech Corn Exchange, built in 1847.

Left: Spalding Corn Exchange, built in 1855, demolished in the 1960s.

Ely Corn Exchange, built in 1847, demolised in 1965.

Spalding Corn Exchange on a Tuesday in the 1950s.

March Corn Exchange, built in 1900.

Britain, however, had repealed the Corn Laws in 1845 allowing hard wheat to be blended with our soft English wheat for bread. In 1877 the price of our wheat had held up fairly well at 52s 9d a quarter and many estate owners thought the repeal of the Corn Laws was not going to affect British agriculture. The Duke of Bedford invested heavily in new farmhouses, buildings and cottages on his Thorney Estate during the nineteenth century.

The price of wheat in 1878 fell to 46s 5d a quarter (£20.25 per ton) and kept falling to 31s a quarter (£13.64 per ton) in 1881. Until well into the twentieth century, cereals were sold by a bushel measure called a 'quarter'. A quarter of wheat was 18 stones, barley 16 stones and oats 12 stones.

From the *Lincolnshire, Boston & Spalding Free Press*, April 1909. The yearly average of wheat has not reached 50s since 1877 when it was 56s 9d. Since that date these are the prices:

1878	46s 5d	1888	31s 10d	1898	34s
1879	43s 10d	1889	29s 9d	1899	25s 8d
1880	44s 4d	1890	29s 9d	1900	28s 11d
1881	31s	1891	37s	1901	26s 9d
1882	45s 1d	1892	30s 3d	1902	28s 1d
1883	41s 7d	1893	26s 4d	1903	26s 9d
1884	35s 8d	1894	22s 10d	1904	28s 4d
1885	32s 10d	1895	23s 1d	1905	29s 8d
1886	32s	1896	26s 2d	1906	28s 3d
1887	32s 6d	1897	30s 2d	1907	30s 7d

The price of wheat after the Second World War remained below £30 per ton until Britain joined the Common Market in 1972 after which prices rose steadily, adjusting to world markets. The price of breadmaking wheat per tonne in 1970/1 was £27.11, 1973/4 £59.85, 1983/4 £135.90 and 1993/4 £123.20, but afterwards dropping in 1989/90 to £86.90. The global market for wheat after that period has been driven more by speculators than physical markets and in 2007 doubled in price, owing to poor predicted harvests around the globe and more cereals destined for biofuels. Since that date we are led to believe that stocks have been replenished and cereals dropped to their 1980s prices.

MILLS

Up until the nineteenth century, cereals were mostly ground or milled by windmills in the Fens and by watermills on the fen edges where flowing water was readily available. Provender mills driven by steam took over from windmills both in villages and on

Two Massey Harris 726 tanker combine harvesters working on J.W.E. Banks' farm, Postland, Crowland, in the 1950s.

A Claas combine harvester combining wheat on reclaimed land near Butterwick, Boston, on Archie Saul's farm in 2008. The sea bank is in the background. *(RS)*

John Deere tractor drilling OSR with Vaderstad 6m minimum tillage drill into wheat stubble in 2007 at the author's farm, Thorney. *(RS)*

large farms in the late nineteenth century. Also at this time, grain could be transported anywhere in the kingdom from farm to the ports or mills by rail.

Post mills and smock mills driven by wind were used to grind the farmers' cereals and pulses up until the late eighteenth century. Most were replaced with tower mills built in the early nineteenth century, many of which have survived to this day. The Fens had its fair share of windmills, many of which still remain. Surviving eighteenth-century mills can be seen at Cowbit, built in 1789, and Lutton, built in 1779. There were also many tower mills and some remaining smock mills at this period which had been used for pumping water which were converted to grinding corn.

Most of the watermills were situated on the fen edge, not in the Fens, on fast-flowing rivers bringing water from the higher ground into the Fens. Indeed, several still remain on the River Welland between Deeping and Stamford. There was a busy lighter trade along this river both ways. One of my ancestors had lighters on this river between Spalding and Stamford in the nineteenth century.

One family synonymous with milling in the Holbeach area is the Biggadike family. The village innkeeper's son, John Thomas Biggadike, from the Saracen's Head Inn at Scaracen's Head near Holbeach worked his way to success, first as a farm labourer when farming was at its lowest ebb for centuries, then as a baker's apprentice at Fletton. His thrifty upbringing enabled him to save enough money and buy Whaplode Mill and bakery in 1890. He ran a very successful business enabling him to retire in about 1917 when he sold the mill and bakery to Sinclair's of Boston. His two sons John and Arthur were away fighting in Europe during the First World War and the chances of their return was slim, which probably tempted him to take the money he had made and put it into good fenland. His son, John Thomas Biggadike 2nd, did survive the war and decided to follow in his father's footsteps as a miller. Whaplode Mill was owned and built by the King family in 1826 who had leased it to the Tindall family of Holbeach. The lease had ended on the mill so John Thomas 2nd took it on in 1924 at £75 per annum. He was only there for four years before the roof blew off, which the landlord replaced.

There had been a smock mill on this site dating back many years which was replaced with the tower mill we see today. The mill has been restored to its former glory having nine floors and is reputedly the tallest in the land. For most of its life it has been without sails which were blown off in a storm in 1884 after which it was powered by a steam engine.

When John Thomas 2nd took the lease in 1924 he installed a Ruston diesel engine which remained its power unit until it was electrified in 1958. The family only ground corn for animal feed, most of which was brought to the mill by the farmer who then collected it after grinding. Oats were rolled mainly for horses up until the advent of tractors and motor vehicles around the time of the Second World War. Competition was fierce at that time with twelve mills competing for trade within a radius of 6 miles of their mill. John Thomas supplemented his income by buying and selling grains and pulses as well as dealing in cattle. Much of his business was done through the Spalding Corn Exchange, both buying and selling. Corn would be bought direct from some of the smaller farmers 'on farm' as well as in the cattle market and corn exchange. They did attend Peterborough Corn Exchange on a shared stand in post-First World War years and later a 'walking on stand' which allowed you to trade without a stand. These stands were only a fraction of the price of a fixed stand.

He would sell to major millers and feed compounders from as far away as Leicester, Lincoln and Manchester, and many more besides. Up to the post-war years these larger firms preferred to trade through local merchants like John who they knew and trusted. Trade for farmers, merchants, millers and livestock dealers in those days was about personal contact and trust. Livestock, cereals, pulses and fertilisers were bought and sold with a shake of the hand sealing the deal – hardly ever was it put into print. A bad reputation was a recipe for disaster.

As the world moved into an era of telephones, motor cars and 'reps on the road', together with the lorry replacing rail transport, the end was in sight for the corn exchanges. Links were severed in the chain of commerce with the larger corn buyers buying direct from the farmer and so the demise of the smaller corn merchant. The national stock feed compounders also had reps on the road selling directly to the farmer. It was a time of change, especially for the 'smaller man' who were John's main customers. John recalls 'the 40-acre farmer'; a 10-, 20- or 30-acre man had no 'clout', but a 40-acre man could, if needs be, throw his weight about. However, his time would come soon.

John traded until 1995, but then closed down the mill after almost seventy-five years of the family's involvement. It has now been restored, except for the sails, by a team of dedicated enthusiasts who have turned it into a major tourist attraction. Hopefully its sails may one day also be restored and the sound of the sails turning in the wind will once again be heard in the village, a sound John Biggadike never heard. It is now believed to be the tallest tower mill in the land still with 8 floors, which at its peak would have over 50 tons of grain stored on them. John would never have wanted to do any other work; his greatest love was meeting people in the 'trade' and farming folk. He does, however, think he should have sold out when his business was in its heyday, as many small millers did at the time. Many millers who expanded their business to survive themselves were eventually processed by the onset of takeovers.

THE PLOWMAN FAMILY

In 1849 Mr George Plowman, the eighth of ten children, set up in business as a miller and baker operating from a post mill in Moulton Chapel. He replaced the mill in 1865 with the present tower mill which his son (George William) took over and ran until he sold it in 1822. Previously G.W. had bought South Holland Mills in Spalding in 1908 from Chamberlin & Co., the lamb food manufacturers. Edgar Plowman, the son of G.W., also followed in his father's footsteps and joined the business in 1908. They managed to survive the great depression of the late 1920s and early 1930s. These were difficult times for the farming industry when many of them found credit more forthcoming from the likes of the Plowmans than their own banks. Without this credit on trust, many more farmers would have gone under. The fourth in line, Edgar's son George Belfitt Plowman, came into the business in 1938 but before he could put his hand to the helm he had to be blooded in war! He saw active service in North Africa and the Italian Campaign, serving with the Derbyshire Yeomanry, a tank regiment. George B was mentioned in dispatches and awarded the MC. He returned home in 1946 to a mill which had been gutted by fire in 1942 when the steam engine used to power the mill had caught fire. This, coupled with the rationing of foodstuffs during the war, did not make good reading for the business. However, George B fought with the same vigour at home as he did in the war, setting about expanding and modernising the company, both on the milling side and grain trading. Mr J.A. Neate joined the grain trading side becoming a director in 1955, regularly attending their own stand in Spalding Corn Exchange. To enhance their service to their farmer customers he attended the corn exchanges in Mark Lane in London, Lincoln and Manchester, making themselves major players in their field. High-quality animal feed was their speciality, especially for pigs, using their own slogan 'Pigs Pay the Plowman Way'. Expansion for the company never ceased by the Plowman family but eventually this policy was not sufficient for them to survive in the larger market place at that time.

George B sold the company including the mill in 1980 to W.J. Oldacre Ltd which closed down in 1998, by then part of an even larger organisation, Dalgety Feeds Ltd. The mill remains, having been converted to flats, many of its occupants probably never knowing of the Plowman family who served local farmers for four generations. George passed away in November 2009.

During the post-Second World War years most towns and a few villages still had a local corn merchant with representatives who regularly called on farms. Many went by the wayside during the 1960s when co-operatives were formed with the assistance of government grants, and merchants were swallowed up in takeovers.

Today we have only a handful of grain merchants remaining. The largest ones are either owned by multi-national companies or have amalgamated with the grain co-ops. There are small family-owned businesses who have survived and one large operator in family hands, Dalmark of Eye. We also have some individuals who trade in grain in localised areas. Three large millers of flour are also present in the Fens as well as the Maud Foster Windmill at Boston who produce stoneground flour.

In the main, cereals are dried and stored on the farm. Some farmers, however, store in co-operative stores, many of whom practice intensive farming and look at cereals as a break crop and do not invest in on farm facilities for handling grain.

Here is a list of merchants and millers still working in the year 2010.

Merchants:
H. & C. Beart Ltd, Brighton Mill, Stowbridge, Downham Market

Charles Wright, Old Leake, Boston

G. Housam & Son, Old Leake, Boston

Dobney, Gosberton Risegate, Spalding

Robert Bellairs from Woodnewton near Oundle trades in the Peterborough and Newborough Fen area

Dalmark of Eye, Peterborough, own the old Dixon-Spain Silo at Thorney Toll

Frontier is an amalgamation of Allied Grain, Cargill and Sidney C. Banks who over the years absorbed the old Dennick Group from Wisbech

Joe Odam from Eye with interests in Gleadalls and Wells Agriculture from Nottingham

Grain Farmers, was originally a tie-up between Dalgety and SCATS, a southern England trading group now trading as Openfield, their head office being at Colsterworth in the Old Viking Grain head office and storage facility

Centaur is the trading arm of Lynn Grain which is also under the Openfields banner
Fengrain, the farmers' co-op from March also trade on the open market
Wellgrain, who are based at Ely, are now run by previous traders of the old Anglia
    Agricultural Merchants business which ceased trading a number of years ago
Cambgrain

Millers:
Oldacres, Holbeach
Whitworths, Peterborough
Favour Parker, Chetisham Silos, Lynn Road, Littleport
Heygates, Downham Market
Maud Foster Mill, Boston

Most of the heavier fen soils are devoted to combinable crops such as wheat, barley, oil seed rape, beans and peas. Some also grow linseed. These farms are, in the main, family farms where all work is carried out by themselves. Some very large units do hire in labour during harvest time, or when required.

Not many farms on this system have survived with less than 500 to 1,000 acres due to economies of scale, and if present economics prevail, they may have to become even larger to survive. John Biggadike remembers some 60 years ago 'the 40-acre farmer with clout'; however, today even the 1,000-acre man lacks clout. This is also reflected in the Crown Estate and county council holdings, once the foundation of such 40-acre men, where few holdings are now less than 500 acres.

One family grain merchant which has survived in this trade since 1870 is G. Housam & Son, Old Leake Mill, Commonside, Boston. Alan runs the grain merchants business and his wife manages the shop which sells a range of pet foods. Standing in their yard is the mill which was built in 1859 and bought by his great-grandfather George Housam in 1870. It has provided for five generations of the Housam family from George, Sam, George, and Alan's son David, and today the mill is still a working part of the business where cereals are milled and rolled for livestock.

Alan buys grain at harvest time mostly from vegetable farmers who have no storage facilities on the farm for the crop. Much of this grain is 'non-assured', meaning it comes from farms who receive the single farm payment but are not crop assured registered, and for this reason the grain is sent for export and not used in the UK. This market is slowly declining as virtually all farms are becoming crop assured.

The shop has changed with the demise of the small farmer who bought chicken and pig feed as well as for sheep and cattle, although most of their food sales are for horses and hobby farmers with various livestock. An astute eye for business and being

Wheat being loaded
into a ship at Sutton
Bridge Docks in 2009.

able to adapt to his marketplace have enabled him and his wife to survive in a very competitive industry, his worst moment being when a large local grain merchant went into receivership owing him money. When questioned about the future, Alan's reply was, 'I am an eternal optimist.' Long may he remain so.

CARROTS

The town of Chatteris became the fenland centre for carrot growing in the nineteenth century. In 1870 over 50 per cent of the town of Chatteris was employed in carrot growing. The town also had its fair share of innovators and entrepreneurs, not only for growing, but harvesting and washing carrots too. Charles Cole of Chatteris developed a mechanical carrot harvester, while innovative washing plants were also being built.

Carrots were extensively grown in that area and the south-east of the Fens. This part of the Fens consisted of mainly deep black fen soil, ideal for growing this root vegetable. Since its reclamation in the seventeenth and eighteenth centuries, shrinkage, burning, paring and surface erosion had already reduced the depth of peat. In the late nineteenth century and up to the mid-twentieth century there was still a considerable depth of peat lying above deposits of marine estuarine muds and gravels. The soil being open and highly organic allowed the carrots to become deeply rooted in the peat, yielding a high tonnage to the acre. Carrots were drilled in beds, and before the winter set in they would be ridged up to cover them with the peat soil to protect them from frost damage. The free-draining soil allowed lifting in most conditions. Gradually, lifting and storage over the winter months replaced leaving them in the field over winter. Storage consisted of various methods to suit each individual farmer. Some were put into small graves, pits or pies, covered over with straw followed by a covering of soil to protect them from frost. In the 1960s the aforementioned 'Dickie Pie' was used, which consisted of straw bales stacked to form rectangular pits. Carrots would be elevated in and covered over with straw and polythene to keep out the weather and the frost. Mr Dickie was a farm manager on a Holbeach Marsh Estate who first used these pies.

Barrel riddling machines were used in the early days to clean, grade and bag the crop. Later it was taken in store where it was graded and sometimes washed. It was nothing unusual to see carrot washing machines alongside a fen drain on the smaller farms, where water was pumped from the drain into the washer. Peat land, by its very nature, contained an abundance of weed seeds and one of the problems with growing carrots on the black peat was controlling weeds. Herbicides for weed control were not available,

Men washing and bagging carrots in a field using the Cole carrot washer, in the Chatteris area in the 1940s. *(Chatteris Museum)*

Carrots being graded and bagged in the field in the Chatteris area in the early twentieth century. *(Chatteris Museum)*

Horse-drawn cart with a load of carrots belonging to J. Boyd of Chatteris. *(Chatteris Museum)*

all control was by hand-weeding. Mechanical hoes pulled by horses and later tractors were in use after the Second World War. The early tractors were mostly run on tractor vaporising oil commonly known as 'paraffin'. Charles Joyce noticed a tractor used for steerage hoeing had leaked fuel onto the carrot crop, which was in its young stages of growth. Some days later he returned to the area and noticed the weeds where the fuel had leaked had died but the carrots were still alive and healthy. From this moment it became the norm to spray 50 gallons of paraffin to the acre on carrots during their early stages of growth. This method continued until the advent of the herbicide Linuron came on to the market in the early 1960s.

The Second World War was the pinnacle of the carrot industry. With their 'Dig for Victory' campaign, the government at the time encouraged the public to eat carrots. Carrots are high in beta carotene so for this reason it was said that they improved night vision during the long periods of the blackout. Most of the bombing missions into Europe were carried out at night time and a reported reason for our success rate was the extraordinary night vision our air crews had through eating carrots! The RAF's top night fighter ace was Flight Lieutenant John Cunningham who became known as 'Cat's Eyes' Cunningham, with twenty German aircraft kills to his name. His fame was founded on his night vision abilities through a love of carrots. The truth was that he was using the newly developed Airborne Interception Radar to home in on his prey, as were some of the bomber crews, which had to be kept secret for the Germans. For this reason alone the carrot myth was high on our propaganda list, as well as earning the growers a good return. This lucrative demand was also the reason the crop was grown at close rotation and infested most of the best carrot land with carrot pests and other diseases. Arthur Rickwood, who farmed 8,000 acres of the black fen, growing 1,500 acres of carrots, was noted by his peers for being the largest carrot grower in the UK (King George VI once called him 'the carrot king'), but little did he realise that by close rotation his land would become infested with eelworm.

By the 1960s and '70s disease-free land for growing carrots was hard to find on the black fen, and black land carrots were not what the pre-packers wanted. This coincided with the growth of the supermarkets who required large volumes of washed, pre-packed carrots of uniform size and shape. A new variety of hybrid carrots called Nantes came on to the market which fitted this demand.

Many growers suffered from poor prices during the 1960s and '70s and ceased growing the crop, and those which remained moved out of the Fens on to areas of light silt land across the Fens. This demand for pre-packed carrots was also being taken up by the Dutch with thousands of tonnes coming from Holland to the UK. The demand by the canners for carrots also grew after the war, reaching its peak in about 1963. They also required small carrots which favoured the sand and silt soils rather than the fen peats.

Chatteris, however, was still a dominant force in the carrot industry and in the 1960s there were still carrot washers in its neighbourhood such as Sid Casey, Munsey, Heading, Alpress, Graves, Rickwood, Barnes, John Sailsbury. E.S. Kemp and Buddles.

Some of the larger growers now looked to the sand lands out of the Fens to grow carrots suitable for pre-packing. Most of these soils contained an abundance of stones which made harvesting and grading difficult. The advent of de-stoning machines and drilling into beds along with irrigation made some Norfolk land, which was previously unsuitable for these crops, ideal carrot-cropping land. There were several factors which changed our growing areas in the UK from the black fens. Irrigation on sand land from deep bore holes changed poor land to fertile soils and new agrochemicals, fertilisers and mechanisation advanced rapidly.

One of the main benefits the deep fen peat soils had was that the carrots could be hilled up with soil for the winter months to protect them from frost. There were also hazards to growing crops on the black fen especially when 'fen blows' occur. One farmer on black fen soil remembers losing 100 acres of young carrots in three days on his farm during one such blow in the spring. Now carrots grown on sand lands in

Carrots being drilled in four-row beds on silt land near Long Sutton in 2010. *(RS)*

Carrot drilling operations of Chatteris in 2010. The machines in the background are steered by Global Positioning Satellite linked to this portable station. *(RS)*

Carrots being harvested during the winter by Albert Bartlett & Sons in 2010. The straw which has protected the crop against frost is being removed by the rear tractor prior to harvesting. (RS)

beds could be covered with straw and polythene by machine to protect them during the winter months. The harvesting period was now extended to enable the growers to supply supermarkets twelve months of the year from UK farms. Sand and light silt soils are, like the black fen, prone to blowing but when the crop is covered with straw this is not a problem. Straw is baled behind the combines mostly in the surrounding Fens and stored in central depot, before being taken from there to the fields where it is required when strawing the crop for the winter. All this does not come cheap – when the crop is neatly bedded down for the winter they have cost in the region of £1,500 per acre.

Apart from a few growers who had managed their rotation to enable them to crop on disease-free land, the carrot crop disappeared from many black fen areas. Carrot growing in the Fens, more than any other vegetable crop, changed its grower status due to the supermarkets' demand for washed carrots. The sand and silt soils produce a long, smooth-skinned carrot as opposed to the shorter, rougher-skinned black land carrot. It is debatable, however, which has the best flavour. I believe the carrot grown on black fen has a sweeter flavour. Organic soils are also prone to carrot diseases more than the silts and sands.

For the first time in several years, Albert Bartlett & Sons are growing carrots on fen peat, and while Chatteris may not be the hub of the carrot growing industry anymore, with Albert Bartlett & Sons in the town, carrots are prepared there for the supermarket. Chatteris and its fringes in the 1960s was the home of ten carrot washing plants, now it has one.

Albert Bartlett & Sons were carrot growers and packers in Airdrie, Scotland, and in 1984 they bought the firm of Barnes in Chatteris. Since that date the firm has expanded the washing and processing factory at Chatteris which is now the most up-to-date in the UK. Total acreage grown for the Chatteris factory is 5,000 acres of carrots, 3,000 of parsnips and 1,500 acres of onions. 1,000 acres of the carrots are grown on the fen silts soils, and 500 acres on black fen. This is one crop in the UK which is dominated by

Carrots after lifting are transported by lorry to Bartlett's factory at Chatteris to be washed, graded and packed.

a handful of large growers, who between them grow around 40,000 acres nationally.

It is still prone to pests, diseases and disorders today, the most problematic being Carrot Fly, Willow Carrot Aphid, Cutworms, Violet Root Rot, Carrot Motley, Dwarf Virus and Cavity Spot. Thanks to modern agrochemicals these problems can be overcome without any health risk to the consumer. The most common varieties grown are Chantenay, Autumn King and Nantes.

VEGETABLES

The growing, processing and marketing of vegetables witnessed untold changes from the 1950s to the present day. The frozen food industry had continued to develop during the Second World War in the USA but not in the UK – our convenience vegetable market was dominated by the canning industry until the 1950s. It was the era of the home deep freezer and to fill them, existing freezing companies expanded their freezing lines as well as new companies coming into the market. The 1960s through to the 1970s saw the frozen food industry for vegetables reach a new peak, with names such as Findus, Smedleys, Birds Eye and Ross being market leaders, alongside newly formed co-ops. This also had to change when supermarkets demanded 'own brand labelling'.

The frozen pea industry had turned their harvesting time from pea viner to the freezing lines to 150 minutes, coining the phrase '150-minute peas'. Peas were and still are one of the best vegetables for freezing but the whole concept of freezing was maintaining quality, 'you can be sure it's fresh if it's frozen' being a well-known frozen foods advertising slogan.

With entry into the Common Market the government of the day poured money into farming co-operatives. Co-ops were formed for growing, harvesting and marketing agricultural produce, although few survived. It was also the era of expanding supermarkets who had fresh vegetables on their agenda. The word supermarket was quickly becoming the most used word in the consumer's vocabulary, a word which would change our entire growing and marketing strategy across the UK. The veg industry would witness more changes over the next two decades than it had in its entire history. Supermarkets requiring frozen as well as pre-packed veg changed their purchasing from the traditional wholesale markets to merchant/growers and to the new farmers' co-ops growing up. Their buyers came to the fen country, many of whom had never heard of the Fens let alone set foot in them. They were joined by plant breeders, plant propagators, machinery manufactures, packaging companies, hauliers, agrochemical companies and even building companies to build temperature-controlled stores. A whole new industry had grown up on the back of vegetable growing in the Fens. It also drew in newly trained people such as technicians, food hygienists and marketing and processing men and women. Of all the food crops grown in the Fens, fresh vegetables have probably seen more changes since the Second World War than any other crop.

TRANSPORTATION

Transportation from the field to the consumer has had its effect on all crops produced here, firstly the railways then road haulage brought the Fens within marketing distance of the centres of mass population. This change started in the mid-nineteenth century lasting until the 1950s to be superseded by road transport. Growing, harvesting, handling and agronomy have also had their effect on all crops grown here. The most striking change in fresh vegetables has been the development in maintaining freshness from the field to the consumer.

Lorries belonging to Lambert Brothers of Donnington loaded with produce for the city markets in the mid-twentieth century.

In the days of rail transport, vegetables would be loaded onto wagons in the local stations for despatch to cities around the country. This could take several hours or days depending on the destination with produce mostly wilted on arrival. Road transport after the Second World War reduced the time factor from farm to market, usually an overnight journey. During the post-war years cauliflowers were cut in the field and thrown into windrows to be then packed into nets, loaded onto trailers and carted out of the field. The nets were then loaded on to flatbed lorries, sheeted down and roped to keep the load secure. Lorries would leave in the evening to arrive in the London markets such as Covent Garden, Brentford and Spitalfields ready for the markets to open in the very early hours of the morning. I also remember those years when streams of lorries heading for the major cities around the UK were seen every evening on the roads out of the Fens. They would be seen again returning empty to their depots in the early mornings, the drivers then going to bed ready for another evening journey. It was no problem then to hitch a ride down to London and back.

The merchants' lorries have changed along with other changes in the cycle. Flat bed lorries were the first but were then replaced by curtain-sided lorries. Artics took over from the flatbeds and eventually temperature-controlled articulated lorries of immense size came into use. Drivers no longer have to drive through the night as most deliveries are carried out during the daytime. Everything is palletised for easy, safe and damage-free handling. Many packers of fresh foods operate their own transport fleets, alongside specialists companies who offer this service.

Returnable crates were introduced after the Second World War when vegetables were cut and put straight into these crates in the field, reducing handling and damage. The problem with these crates was the cost of hire and the labour involved in returning them to the farms for refilling. The crates mainly belonged to the city merchants which restricted who had your produce. These were followed by non-returnable crates which reduced handling but were still filled in the field. Many growers then reverted to cutting into large crates mounted on tractors which were taken to a central pack house.

Harvesting Brussels sprouts on Sam Cooper's farm at Bicker, for Clements of Wrangle, in 2009. *(RS)*

Spalding Bulb & Produce Auction in 2009. Much of the produce from small growers is sent in recycled boxes.

Below, right: Willie Chappel of Gosberton Fen cutting cabbages bound for the Produce Auction at Spalding in recycled boxes in 2009. *(RS)*

Packing rigs came into operation in the 1960s designed to travel in the fields with a gang of cutters feeding the mobile pack house.

The cut veg is put on to endless conveyors with cups to hold the cauliflowers/cabbage which feed into the rig where it is packed into crates or overwrapped with clingfilm. From the field it is transported to the merchants' central pack houses. All veg is harvested this way by rigs but some veg is taken back to the central pack houses for further packing or processing, such as broccoli, or prepared for freezing companies.

Almost everyone handling large quantities of these vegetables use rigs and brussels sprout harvesters today. Some sprouts are harvested by hand but on a limited scale – Marks & Spencer do sell hand-picked sprouts.

PLANT RAISING

Other major changes to vegetable production also took place in the field. The traditional method of growing most brassica crops was the transplanting of plants from the plant raiser to the field. The seed was either sown under glass for early plants or outside for later plants. When the plants were ready for transplanting they were pulled by hand and put into trays to be taken to the fields for planting, usually in rows with 18- or 20in spacing. Problems could occur when soil conditions were too dry and the plant struggled to establish itself – conversely, if conditions became too wet, the plants would wilt and lose their vigour, or even die before planting. This method went through a revolution in the late 1970s with the introduction of modules.

Planting vegetable plants in the mid-twentieth century.

Seeds are now drilled into trays of modules by specialist plant raisers, then grown in glass houses. While in trays the plants can be controlled in their growth stage to coincide with planting. When needed for planting, the trays of modules are sent to the fields where they are transplanted into the soil by mechanical planters operated by workers. This module process enables more precise planting programmes and if conditions are not suitable for transplanting, they can be held in cold stores until a better time.

The size of the module the seed is planted in can also determine the time of harvest date – the larger the module used, the earlier it will mature. It has also enabled the grower to achieve a more even growth in his crops and so he can harvest in less operations. A crop of cauliflowers can be harvested by going through it four times, where as in the past it may have taken double that number of cuttings. Meanwhile, cabbage, now, is often harvested by going through the crop twice. Crops are now in the soil less time than they were in the past allowing more timeliness for field operations.

To make operations easier when going through the growing crop, tramlines (a gap in the crop the width of a tractor) are left fallow. This is then used for applying agrochemicals, fertilisers and for harvesting rigs to travel along. Larger flotation tyres are a feature of modern machines to avoid soil compaction and with wider sprayers and fertiliser spreaders, the width between tramlines has increased. Even with the vast changes in the veg growing business today, some things have not changed. Each row of cauliflowers sees the human foot several times at harvesting and the cutter's knife is still the same as it was when cauliflowers were put into returnable crates in the 1950s.

For early veg, modules are planted in the field and then covered with polythene covers to encourage early maturing, extending the home-produced crop's window.

FRESHNESS

To keep veg fresh the field heat must be taken out of it as quickly as possible. This may not be so critical during the winter months, especially when the north-east winds blow across the bleak marshes, but it is during the other seasons of the year. As soon as it arrives at the

Salad cabbage going into controlled environment store for Merry Mac of Manea in 2010.

pack house if there is not further preparation needed it goes directly in to a cold store. From here on, vegetables remain at a constant temperature, in transport and storage mode until presented on the supermarkets' shelves. So precise are these stages in the process of conveying the veg from field to the supermarket shelf that, even when loaded and unloaded in and out of the fridge vans, care is taken with sealed doors to maintain a cold temperature.

The general public have never in history had the choice and quality of vegetables and fruit they have today. This care in handling the crops has boosted the sales of fresh veg in the UK because it maintains its freshness, probably at the expense of frozen veg.

PLANT BREEDING

In the early days cauliflowers were either autumn cauliflowers or overwintered broccoli (also known cauliflower), leaving certain periods of the year in the UK with no

Clements of Wrangle's cauliflower harvesting rigs in 2009. *(RS)*

cauliflowers, relying on imports. The introduction of the summer cauliflower in the 1960s changed the growers' harvesting patterns. Advances have been made in the plant breeding of all veg and now growers harvest cauliflower and cabbage twelve months of the year. The varieties harvested during the winter months today have the outer leaves covering the crown protecting it against the elements and other pests which could damage them before harvesting. Calabrese was introduced in the 1960s and quickly established itself as major part of the veg market and at one time was a serious contender to the cauliflower.

Like other agricultural crops the vegetable has progressed with plant breeding. The seed before the hybrids were introduced were open pollinated where quality was not always reliable. Now all seed from the seed houses are hybrid varieties.

Chemicals in conjunction with plant breeding play an important part in crop science of today, not just in pest control during growing, but in seed dressing. Precise applications of sophisticated chemicals are available which the plant can take up to protect the micro-environment around the new emerging seeds, as well as chemicals to be taken up systematically to protect the plants in their early stages.

The day will come when genetically modified plant breeding, like hybrids, will be available across the whole world, not just a few countries, if mankind is to feed himself with the limited resources he has at hand. To reduce our reliance on agrochemicals could be beneficial to mankind and the environment. The controversy over this issue is not about benefits of GM but a public relations exercise cleverly manipulated by the media and the anti-GM lobby.

Clements of Wrangle is the name seen on many large temperature-controlled articulated vehicles, heading along the motorways of the UK. Many people seeing them have no idea what is inside them when they have left the Lincolnshire Fens, or even where their destination will be. Inside those vehicles is fresh produce grown on 5,000 acres, much of which is on the finest land in the Fens. Their cargo includes pointed and green winter cabbages in spring, cauliflowers twelve months of the year, broccoli in season, and brussels sprouts for the winter months. It is a streamlined hi-tech operation to grow, harvest and deliver quality vegetables as fresh as the day they are cut in the fields. Alan Clements and his son Christopher supervise the operations of the business. The family started from humble beginnings; Alan's grandfather kept a pub while his father

Cauliflowers being packed in harvesting rig for Clements. *(RS)*

started his working life on a farm owned by Fred Grant at Freiston near Boston. He decided to apply to the Lincolnshire County Council for one of their small-holdings near Eastville between the two world wars. His application was successful and he moved in to a holding of 20 acres. He worked hard and prospered and, as a result, in 1945 he bought 15 acres with a house and premises near Wrangle. The soil is extremely fertile around Wrangle and is perfect land for growing cauliflowers and very soon afterwards he rented 7 acres close by, also good vegetable growing soil. Commenting on the LCC's policy today of offering vacant holdings to neighbouring tenants, Alan's comments were, 'It does make existing holdings more viable but restricts new entrants coming into farming.' Words of experience.

The intention of the Small-holdings Act of 1908 was for county councils to make available land 'for a full-time opportunity for entry into commercial agriculture and horticulture.' The Clements family are without doubt one of the greatest success stories of this scheme in the Fens, if not the entire country, where county council holdings were set up. Without that start on the bottom rung of the ladder they may not be where they are today. After the Second World War many local potato and vegetable merchants grew up in the Boston area to supply all major cities across the UK. Clements sent their vegetables through some of these merchants and also via local haulage companies to merchants in the wholesale markets. Potatoes were bought by the merchants at an agreed price whereas vegetables were mainly sent to be sold on a commission basis. The grower received his returns after the seller had deducted his commission.

The 1960s was the decade of exciting marketing changes in vegetables. Everyone would have to change and adapt to these marketing forces if they were to survive. One of their first ventures into the food processing industry was growing and supplying sprouts to one of the national freezer companies.

By 1961 Alan and his father had expanded their vegetable acreage and began marketing their own vegetables to the wholesale markets such as London, Nottingham and Leicester. This was the same year they bought their first lorry.

In the 1960s the word supermarket was quickly becoming a household word and would very soon filter down to the Fens, especially to the vegetable growers. Clements were still supplying the wholesale markets well into the 1970s but by 1978 they had established links with the supermarkets. Their business changed dramatically when they started supplying them. The first major change was the packing of vegetables followed by modern handling and harvesting techniques. The next two decades transformed the business from a small family-run operation to a multi-million pound growing and marketing organisation. Today they market vegetable products from 5,700 acres – this includes their own land, land hired, as well as produce from other growers. Their products sold during 2008 include the following:

Baby cauli/broccoli	3,000,000 heads
Broccoli	10,600,000kg
Brussels Sprouts	4,300,000kg
Cabbage	8,000,000 heads
Cauliflower	11,400,000 heads
Purple sprouting broccoli	40,000kg
Ready prepared vegetables	2,500,000kg

Some years ago a break in the rotation from vegetables was not considered essential, but due to a build-up of disease their policy has changed. Vining peas, for Bishop Peas, have been added to the rotation for a break crop as well as being an ideal entry for a vegetable crop in midsummer. Potatoes are also now back in the rotation. Clements, like many other vegetable growers, are very conscious of soil-borne diseases.

A vegetable planter with five operators behind crates of modules feeding them into the machine in 2010. All were Eastern European labourers. *(RS)*

Their supply base covers a wide area – in England they serve Lincolnshire, Suffolk, Norfolk and Cornwall. Cauliflower is grown for them in Spain while spot buying is carried out in France, Poland and Holland for most veg. To be a supplier to the supermarkets, firms like Clements have to be able to supply veg twelve months of the year in all circumstances. To do this their supply base covers a vast area to enable them to source their goods. Like many other suppliers to the supermarkets they also export to Ireland and some parts of Europe when availability and price fit the bill. Freezing companies and prepared-food suppliers are on their customer list as are 'low-risk prepared foods'. This is veg prepared in convenient packs ready for cooking, but not washed.

Everything Clements grows there is an outlet for, in one form or another. This continuity is not easy to achieve on a fixed acreage of cropping. Even with the sophisticated agronomy and programming, the weather still plays a critical part in planting, growing, harvesting, size and quality of the crop. A market can be found for surpluses of veg, but in a time of shortage due to weather, that shortage has to be bought in to fulfil their contracts to their customers. In the past, growers would talk of a 'weather market' when due to bad weather, crops were short in the wholesale markets. This is very rare today.

Twenty years ago all labour in the industry was sourced locally. The demand for prepared foods and packed veg created a void in the labour market which has been filled by Eastern European labour. Clements still employ local labour but most of the workforce working in the fields are from Russia, Poland, Romania and Latvia. English labourers prefer to work indoors in pack houses and prepared food factories. All their casual labour for field operations are supplied through local licensed gangmasters who source their labour from the UK and Eastern Europe. Alan insists that the veg industry could not survive without Eastern European labour and has a high regard for them.

To maintain the quality demanded today, for supermarkets, freezers and processors they are always trying to reduce wastage of the crop to achieve these standards. The price structure has changed since the days of the wholesale markets. Gone are the commission days, not knowing what your veg came to when it left the farm. Most supermarkets work on a weekly price mutually agreed, but determined by them. The food services industry buy on a seasonal price whereas the freezers buy on an annual wholesale price. Like many of his competitors in the veg business, by far the main part of his business is with a very few supermarkets whose demands are stringent but their cheque books are safe. Changes are always around the corner. Many years ago a rotation was essential to control soil-borne diseases. This changed to more of a monocultural system of growing veg which resulted in disease problems in the soil. Now growers such as Alan are more concerned for the long-term future of their soils and have gone back to a wider rotation for veg. The cultivation methods are also changing by reducing heavier field operations such as ploughing in some instances and some power harrow operations which will all help the structure of their soils.

While writing this book the family have just bought a neighbouring farm to add to their existing holding. The right soil for the right crops is still the most important factor in the industry, and they don't make any more of it! Alan would not tell how much he

Harvesting leeks on Allpress Farms, Chatteris, 2009. Leeks are cut, fed into the machine where they are trimmed for dispatch to pack house. Left to right are Lenton Allpress and his son Patrick. *(RS)*

Inside the Allpress leek harvesting rig in 2009. *(RS)*

Inside the Allpress pack house, each leek is weighed automatically and graded to enable numbers in each pack to be the correct weight.

paid for the farm, but I know that it must have been another record-breaking price for some of the finest soil in the land, and always will be.

Allpress Farms of Chatteris has a track record of four generations farming in this locality, growing wheat, sugar beet, onions and leeks on 1,660 acres. They produce 7,500 tonnes of leeks and 5,000 tonnes of onions per annum mostly for Sainsbury's, who they have supplied with leeks since 1970. The present generation at the cutting-edge are Nick and Patrick with their father Lenton still keeping a watchful eye over the business. Sainsbury's take 19 million loose leeks, 25 million extra trimmed leeks and 6 million baby leeks, figures which are hard to contemplate. Their operation is mainly on the peat soils in that area which are suited to harvesting during the winter months when this vegetable is in demand. This is a vegetable growing operation on a large scale, even by fen standards with a high labour input, mostly made up of men and women from Eastern Europe.